maths CHALLENGE

Edited by Tony Gardiner

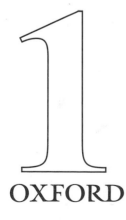

OXFORD

OXFORD
UNIVERSITY PRESS

Great Clarendon Street, Oxford OX2 6DP

Oxford University Press is a department of the University of Oxford. It furthers the University's objective of excellence in research, scholarship, and education by publishing worldwide in

Oxford New York

Athens Auckland Bangkok Bogotá Buenos Aires Calcutta Cape Town Chennai Dar es Salaam Delhi Florence Hong Kong Istanbul Karachi Kuala Lumpur Madrid Melbourne Mexico City Mumbai Nairobi Paris São Paulo Shanghai Singapore Taipei Tokyo Toronto Warsaw

with associated companies in Berlin Ibadan

The materials were written, tested and revised over a period of several years by a group of practising teachers. The main contributors were: Tony Burghall, Peter Critchley, Michael Darby, Tony Gardiner, Bob Hartman, Mark Patmore, Diana Sharvill, Rosemary Tenison. They acknowledge additional contributions from Geoff Dunne and Peter Ransom. The final draft benefitted from scrutiny by Peter Moody. The moral rights of the authors have been asserted.

Throughout the development phase, the group's work was supported by the School Mathematics Project.

Database right Oxford University Press (maker)

First published 2000

British Library Cataloguing in Publication Data. Data available.

ISBN 0 19 914777 9

Typeset in Great Britain by Tradespools Ltd.
Printed in Great Britain by Butler and Tanner

CHALLENGES

COMMENTS & SOLUTIONS

GLOSSARY

About this book

This book is the first in a series of three books covering Years 7–10 aimed at the top 10% or so of pupils aged 11–15. The series has been written and produced so that successive volumes provide an enrichment programme for the most able pupils in the lower secondary school, though many items could be adapted for use with other age groups and with a wider ability range. Each individual section in this book can also be used to form the basis of an excellent 'masterclass' for secondary pupils.

Each book in the series contains twenty or more sections together with Comments and Solutions. These Comments and Solutions provide more than just a convenient way for pupils to check their answers: they have been written to help provoke further thought and discussion with peers or with a teacher. Teachers may have to stress that, even when pupils think they have succeeded in solving a problem, it is important for them to reflect on the approach they have used (and the reason why it works). In particular, they should never be satisfied with merely getting the answer.

The material has been extensively trialled and revised in the light of teachers' comments. The styles of the different sections are very varied. This provides a freshness of approach, which not only adds to the value of the collection, but which challenges pupils to think about familiar topics in new ways. Though different items are written in different styles, all sections seek to challenge able pupils in ways that will help them to:

- develop a deeper understanding of the ideas and methods which are central to secondary mathematics

- lay a firmer foundation for later work

- develop reasoning skills

- cultivate a broader interest in mathematics

- develop their ability to research and to interpret mathematics.

The potential of the material is illustrated by three comments from teachers who trialled one batch of draft material.

'Excellent item on a topic that is rarely understood in depth.'

'If only my A-level students had seen this in Year 7!'

'This item is nicely constructed in a way which makes it suitable for either individuals or small groups working on their own.'

Other materials and strategies which are currently used to extend more able pupils tend either to provide interesting 'time-fillers' which have little to do with ordinary curriculum work, or to accelerate pupils through the National Curriculum levels (and the associated examinations) ahead of their peers. In contrast, this book contains activities which encourage pupils to think more deeply about relatively elementary mainstream material while covering a fairly broad range of standard curriculum topics. The book also includes some extension material which enriches the standard curriculum. Thus it offers schools an alternative both to using time-fillers, and to the acceleration strategy referred to above (a strategy which puts considerable burdens on staff, and which leads to able pupils being permanently out of phase with their peers).

Contents

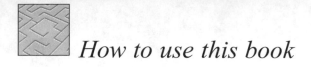

How to use this book

To teachers and parents

• This book is aimed at pupils with the same kind of background as the top 10% of pupils in Years 7–8 in English schools (i.e. pupils aged 11–13). It is the first of a series of three books containing material to challenge and extend such pupils in Years 7–10 (aged 11–15).

• Many of the sections included can be completed by pupils working on their own; the **Comments and Solutions** provide additional support. However, all sections would benefit from some discussion – either with other pupils as part of the activity, or with a teacher or parent.

• A brief summary of the intended learning objectives is given at the beginning of each section, along with the intended classroom organisation (whether individual, or small group work), any necessary materials which must be available, and any sections which should have been tackled before you start.

   This information is provided both to help the experienced teacher, who may be using the material for the first time, and to alert pupils to any necessary prerequisites.

• When pupils do not understand a technical term which appears in the text, they should first consult the **Glossary** at the back of the book.

• The amount of time a given pupil spends on each section will vary. However, we anticipate that each shorter section should take 45–60 minutes, while each longer section may take 1–2 hours.

• Some sections explicitly tell pupils to *choose* a task which interests them, rather than to attempt every single task. Even where pupils are expected to work systematically through a section, it is not essential for them to complete every single question.

How to use this book

To pupils

- These activities challenge you to think more seriously about some of the mathematics you have learned, and about the way you do mathematics. This will not only improve what you can do this year, but will help to lay a solid foundation for future work.

- Some problems are straightforward; other problems are rather hard. Don't give up too easily. If the last couple of questions in a section are too difficult, or involve ideas which you have not yet met, don't worry. It is more important to think carefully about the problems you do than to worry about the ones you leave undone.

- **You are not expected to tackle every single section!** Nor do you have to complete every single question in the sections you do tackle. The material is meant to be fun – as well as being challenging. So if you have to choose, concentrate on those items that interest you most. However, a problem will often not make sense unless you have understood earlier problems in the same section; so **make sure you complete the first few questions in each section** that you tackle.

- Many of the sections can be completed on your own. Some require you to talk to someone else about the ideas involved. It can often help to talk things through with someone else, even when it is not strictly needed in order to complete the task.

- Some sections may require you to visit a library or to use reference books. In addition there is a **Glossary** (that is, a mini-dictionary) at the end of the book which should help to explain any unfamiliar mathematical words or symbols which are used in the text.

- The **Comments and Solutions** are included to help you. Use them wisely! Don't peep too soon; but don't be afraid to use them to help you get unstuck. If you struggle to understand the Comments and Solutions for one question in a section, it will often help you to tackle other questions in that section.

  Remember that the most important thing is not just to get 'the answer. You should always present the method you use in a way that shows that it is correct. In mathematics this often means that a full solution to a problem has to use algebra!

1 BIG numbers

This activity focuses on:

- looking up numerical facts;
- learning to handle large numbers with confidence;
- deciding what calculation to do;
- simplifying calculations by using rounding.

Organisation/Before you start

- You will need a pound coin, a ruler and a map of Britain. You will also need to weigh a pound coin (or estimate its weight).
- If you are working with a partner, pick different questions to work on individually. Then explain your methods to each other.

Do a simple calculation to find the approximate answer to each question.

1 Do human beings live for as long as a million hours?

2 If you have lived for a million seconds, how many birthdays have you had?

3 What year was it one billion seconds after the birth of Christ?

4 What year was it one billion minutes ago?

5 How long would it take to count to a million?

6 If five million one pound coins are laid edge to edge along the M1 starting from Leeds and going south, how far will they reach?

7 How many one pound coins would you need to make a line from Land's End to John O'Groats?

8 If you walk at an average speed of 3 miles an hour along a line of pound coins, how long would it take you to walk past a million pounds worth of coins?

9 Is a pile of one million one pound coins as high as the Eiffel Tower in Paris?

10 Suppose you were worth your weight in one pound coins. How much would you be worth?

11 Would one million one pound coins fit in a telephone box?

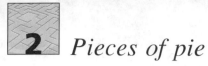

2 Pieces of pie

This activity focusses on:

- using simple mathematics to make sense of a slightly puzzling object;
- performing informal calculations involving simple fractions.

This is a picture of an old 'pie plate' (that is, a plate you put pies on). The plate was made in France. It has strange markings round the edge. The markings have been printed round the outside of the picture so that they are easier to see; they are meant to make it easier to divide the pie fairly.

1   **a** What do you think the numbers round the edge of the plate mean?

b How would the markings help someone to divide a pie fairly into seven pieces?

2   Explain briefly how you could use the plate to show that these pairs of fractions are equivalent:

a $\frac{1}{3}, \frac{2}{6}$

b $\frac{2}{3}, \frac{6}{9}$

c $\frac{2}{6}, \frac{3}{9}$

3   Use the markings on the pie plate to help you calculate these.

a $\frac{1}{2} + \frac{1}{6}$

b $\frac{1}{4} + \frac{3}{8}$

c $\frac{2}{3} + \frac{1}{6}$

d $\frac{1}{2} - \frac{1}{6}$

e $\frac{3}{4} - \frac{1}{8}$

f $\frac{2}{3} - \frac{1}{9}$

4   One way to think about the calculation $10 \div 5$ is to ask:

'How many times will 5 fit into 10?'
The answer is: Two times. So $10 \div 5 = 2$.

Use this way of thinking about division (and the pie plate scale) to work out these.

a $\frac{1}{2} \div \frac{1}{6}$

b $\frac{2}{3} \div \frac{1}{9}$

c $1 \div \frac{1}{7}$

3 Equals

This activity focuses on:

- realising that ' = ' can have different meanings;
- recognising common misuses of ' = ' and learning to avoid them;
- appreciating something of the history of mathematics.

Organisation/Before you start

- For question 5 you need access to dictionaries and an encyclopedia.

The Whetstone of Witte

The first algebra book in English, *The Whetstone of Witte*, was written by Robert Recorde and published in 1557. In this book the equals sign ' = ' was used for the first time. Before this time people wrote 'eq.' as a shorthand for the Latin word for 'is equal to'.

The title page of the book is shown below on the left; the page where the equals sign is introduced is shown on the right.

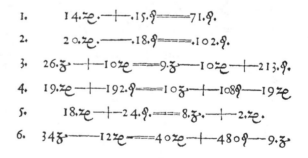

1 Why do you think Robert Recorde chose the title *The Whetstone of Witte*?

2 What did Recorde give as his reasons for choosing the = sign? (The old-fashioned printing is hard to read. Part of the challenge in this question is for you to piece together as much as you need to answer the question. Notice that the letter 's' is printed as 'f', and some of the spelling is strange – for example, 'more' is spelt 'moare'.)

maths CHALLENGE 1   4

Symbols and words

3 Match each symbol to the word phrase it represents.

Symbols:
$=, \approx, \leq, \neq, \geq, \equiv, \simeq$

Phrases:
is not equal to
is identically equal to (or
is equivalent to)
is greater than or equal to
is approximately equal to
is equal to
is less than or equal to

Look it up

4 a Suppose you had to explain the meaning of the symbol ' $=$ ' to a younger brother or sister. What would you say?

b Does the equals sign have the same meaning every time it is used? How many different meanings of the equals sign can you think of? Give examples to show what you mean.

5 a Look up the word 'equals' in a dictionary (or dictionaries) and in an encyclopedia. Make a list of its many (mathematical) meanings.

b Do any of your sources give a clear meaning for the **symbol ' $=$ '** ?

Equations

The equations-in-words below illustrate certain other words or phrases we sometimes use instead of writing a mathematical equation using ' $=$ '.

$4 + 3$ *makes* 7; $4 + 3$ *is equivalent to* 7; $4 + 3$ *is the same as* 7;
$4 + 3$ *is* 7;   7 *is* 3 *more than* 4;   7 *take away* 4 *leaves* 3.

6 What word (or words) might you use instead of ' $=$ ' in these equations?

a $\frac{4}{6} = \frac{2}{3}$

b $4 \times 5 = 5 \times 4$

Equal or not?

The symbol '$=$' is sometimes used incorrectly or inappropriately. Here are some examples. Rewrite each example either in words, or using a different (correct) symbol.

7 1 cup of coffee $=$ 45p

8 Scale 1 cm $=$ 10 m

9 $\pi = 3.14$

10 Henry VIII $=$ Anne Boleyn

11 $= 1000$ fish

12 $2.4807 = 2.5$ (Pupil's answer to: 'Write 2.4807 correct to one decimal place.')

13 $4 + 5 = 9 \times 3 = 27 - 7 = 20$ (Pupil's answer to: 'Work out $(4 + 5) \times 3 - 7$.')

14 $3x + 4 = 19$
$= 3x = 19 - 4$
$= 3x = 15$
$= \ x = 5$

4 *True, False and 'Iffy' number statements*

This activity focuses on:

- thinking logically;
- making statements precise;
- identifying and resolving common number mistakes.

Organisation/Before you start:

- This section is best tackled by two or three pupils working together. (If you are working on your own, imagine how you would convince others that your decisions are correct.)

Some of the statements on the next page are **Definitely true**; others are **Definitely false**; and some are '**Iffy**' (true if . . .).

An '**Iffy**' statement is one that is not true as it stands, but which could become true if it was made more precise.

For example: The statement

Multiples of 5 end in 5

is '**Iffy**', because:

some multiples of 5 do end in 5 (e.g. 15, 25, . . .),
but other multiples of 5 do not end in 5 (e.g. 10, 20, . . .).

---

1 **a** Make copies of the 18 statements on the next page and cut them out so that you have each statement on a separate slip of paper.

b Make a table with four columns headed like this:

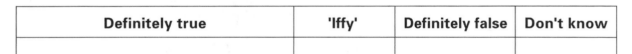

Definitely true	'Iffy'	Definitely false	Don't know

The **Definitely true** column should be about half a page wide.
The other columns should be about one sixth of a page wide.

c Pick one of the statements **A–R**.
Decide which column it should go in and put it there.
Only use the **Don't know** column if you do not understand the statement, or if you cannot decide where it should go.
(**Remember:** 'Iffy' is not the same as **Don't know**.)
Repeat this for each of the statements **A–R**.

d Now pick one of the statements that you have put in the '**Iffy**' column.

Try to make the statement more precise so that your improved version could go in the **Definitely true** column.
For example, to make the statement

Multiples of 5 end in 5

more precise so that it becomes **Definitely true** we could change it to:

Multiples of 5 end in a 5 or a 0,
or   *Odd multiples of 5 always end in 5,*
or   *Any number ending in 5 is a multiple of 5.*

Do this for each statement you put in the '**Iffy**' column.

e Ask your teacher about statements in your **Don't know** column.

A The product of two numbers is a whole number.	B Adding a zero to the end of a number multiplies it by 10.
C The square root of any number is smaller than the original number.	D The product of two odd numbers is odd.
E Every square number has an odd number of factors.	F When you square a number the answer is positive.
G Suppose you divide a number by 2 and then by 10. If instead you divide the number by 10 and then by 2, the answer could be different.	H Calculating two fifths of a number is the same as dividing it by 5, then multiplying the answer by 2.
I Prime numbers are odd.	J The sum of the digits of any multiple of 3 is divisible by 3.
K The sum of two numbers is greater than their difference.	L The only factors of a perfect square are the square itself, its square root, and 1.
M The product of three whole numbers is never the same as their sum.	N The product of a negative number and a positive number is negative.
O Dividing by a number less than 1 gives a larger number.	P When you multiply two numbers the answer is bigger than either of them.
Q The sum of two odd numbers can sometimes be odd.	R The cube of any number is bigger than its square.

5 Möbius bands

This activity focuses on:

- visualising what will happen in geometrical experiments;
- making sense of surprising outcomes.

Organisation/Before you start:

- You will need plain paper, a ruler, scissors, and glue.
- This work is best done by two or three pupils working together.

1 Cut out a strip of paper about 30 cm long and 2 cm wide.

Draw a line down the middle of the strip on both sides of the strip.

Bring the two ends of the strip together as though you intend to make an ordinary loop – but before you glue the two ends together put a **half twist** (180°) in one of the ends.

Then glue the two ends together to make a **Möbius band**. Make sure the ends are stuck firmly together!

a Imagine cutting all the way round – along the line down the middle of the strip. What do you think will happen to your Möbius band when you have finished cutting all the way round?

b Now cut along the centre line to see if your prediction was correct. Write down what actually happens. Can you work out why?

c What do you think will happen if you draw a line down the middle of the new strip and then cut all the way round? Try it!

2 Cut out a new starting strip; mark a line down the middle on both sides as before. Bring the two ends together as though you intend to make an ordinary loop, but this time put a **full twist** (360°) in one end before you glue the ends together.

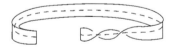

a What do you think will happen to this band when you cut all the way round along the centre line?

b Try it. Was your prediction correct? If not, try to explain what actually happens.

3 Cut out a new strip. This time mark two lines along the whole length of the strip on each side of the strip – each line **one third** of the width from each edge.

Join the two ends with a 180° half twist and glue.

a What do you expect will happen when you cut along one of the lines?

b Now cut and see if you were right.

4 Cut two strips of paper 20 cm long and 2 cm wide.

Draw a line down the centre, on both sides of each strip.

Glue the ends of the strips (without twists) to make two ordinary loops. Then stick the two loops together at right angles as shown.

a What do you expect will happen if you cut along the centre lines of each of the two loops?

b Now cut to see what really happens.
Try to see how you could have predicted what actually happens.

5 Put two loops together with one inside the other.
Then cut along the centre lines.

a What happens when you glue the loops together on one side only?

b What happens when you glue them together on both sides?

c Explore what happens if one or both of the two loops have twists in them.

Glue

Glue on one side.

or

Glue on two sides.

6   *Fractions with, and without, a calculator*

This activity focuses on:

- exploring different ways of deciding which is the larger of two given fractions, and understanding the connections between them;
- practising different ways of calculating with fractions;
- establishing reliable and efficient ways of comparing fractions.

Organisation/Before you start:

- You will need to know how to add and subtract fractions.

CHALLENGES

Which is bigger: $\frac{13}{18}$ or $\frac{11}{15}$?

1   Without using a calculator and without doing any written calculation, decide which of the two numbers $\frac{13}{18}$ and $\frac{11}{15}$ is biggest?

Explain clearly the reasons for your answer.

The following questions explore five different, but related, ways of deciding which of these two (or any two) given fractions is largest. As you work through the questions:

- Check whether your **answer** to question **1** was correct.

- Check whether the **reason** you gave was correct, or how it could have been improved.

With a calculator (if you wish)

Method 1: Subtraction

2   Work out one of the two subtractions: $\frac{13}{18} - \frac{11}{15}$ or $\frac{11}{15} - \frac{13}{18}$.

How does your answer tell you which of the two fractions is biggest?

Method 2: Fraction of a suitably chosen quantity N

3   Choose an integer N that is divisible both by 18 and by 15.

Work out $\frac{13}{18}$ of N. Work out $\frac{11}{15}$ of N.

How do your two answers tell you which of the two fractions is biggest?

Fractions with, and without, a calculator   11

Method 3: Equivalent fractions

4 Produce a list of fractions equivalent to $\frac{11}{15}$:

$$\frac{11}{15} = \frac{22}{30} = \frac{33}{45} = \frac{44}{60} = \ldots\ldots$$

Produce a list of fractions equivalent to $\frac{13}{18}$:

$$\frac{13}{18} = \frac{26}{36} = \frac{39}{54} = \frac{52}{72} = \ldots\ldots$$

Extend both lists until you find two fractions with the same denominator – one in each list. Write down these two fractions.

Compare the numerators of these two fractions.

How does this tell you which of the two fractions $\frac{13}{18}$ and $\frac{11}{15}$ is biggest?

Method 4: Comparing decimal approximations

5 Evaluate each fraction as a decimal.

How do your two answers tell your which fraction is biggest?

6 Look carefully at each of the above four methods.
Make sure that you understand how to carry out each method **without using a calculator**.

Further problems

7 $\frac{2}{3}$ and $\frac{4}{5}$ are familiar fractions.

   a Work out in your head which is the larger.

   b Which method would you use to check your answer using a calculator? Why?

   c Which method would you use without a calculator? Why?

8   a Here is another method.

Method 5: Cross-multiplication

Suppose you are given two fractions (such as $\frac{2}{3}$ and $\frac{4}{5}$).

First multiply the numerator '2' of $\frac{2}{3}$ by the denominator '5' of $\frac{4}{5}$.

Then multiply the denominator '3' of $\frac{2}{3}$ by the numerator '4' of $\frac{4}{5}$.

Compare the two answers. How does this tell you which fraction is the larger of the two? Explain.

b Use Method 5 to compare $\frac{11}{15}$ and $\frac{13}{18}$.

9   How would you work out which is the larger of $\frac{2}{3}$ and $\frac{5}{8}$?

EXTRA 1

Washing up

During the holidays you agree to do the washing up at home for 1p on the first day, 2p on the second day, 4p on the third, and so on, doubling the amount each day. How much would you earn in one week? In two weeks? In three weeks? In four weeks?

Solution: page 77

CHALLENGES

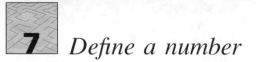

7 *Define a number*

This activity focuses on:

- using mathematical language precisely;
- learning to handle several facts at once.

Organisation/Before you start:

- This work is best done by two or three pupils working together.

Each of these statements could describe a positive integer. Make a copy of all 26 statements for easy reference.

A It is even.	B It is odd.	C It is prime.
D It is a square.	E It is a cube.	F It has one digit.
G It has two digits.	H It is divisible by three.	I Its units digit is 3.
J It is the product of two different primes.	K It is divisible by 6.	L It is divisible by 5.
M It is a multiple of 11.	N The sum of its digits is 10.	O It has a remainder of 1 when divided by 4.
P It is a triangular number.	Q It has a factor (other than 1 and itself) which is a square.	R It is less than 20.
S It is greater than 20.	T It is a multiple of 7.	U The units digit is greater than 5.
V The sum of its digits is 9.	W It is a factor of 60.	X It is the sum of two primes.
Y The units digit is less than 5.	Z The product of its digits is even.	

Make sure you understand what each statement means before trying these questions.

1 a Statements **D**, **G** and **O** all fit the number 25. Which other statements also fit the number 25?

   b Choose another integer between 1 and 50. Find **all** the statements that fit your number. Discuss your answer with your partner(s).

2 **a** Which numbers less than or equal to 30 fit statement **J**?

    b Which statements are true for the number 15?

    c Which numbers less than or equal to 30 fit both statements **G** and **Q**?

    d Which numbers less than or equal to 30 fit all three of the statements **B**, **C** and **S**?

    e Which statements are true for exactly three integers between 1 and 30?

3 **a** I'm thinking of a number which fits both statement **A** and statement **C**. What must my number be?

    b I'm thinking of a number less than one thousand that fits all three of the statements **B**, **D**, **L**. How many different possibilities are there for my number?

    c Play the following game with a partner:

- You each choose a number less than or equal to 30. Don't tell your partner what your number is!

- Each of you then writes down a list of **all** those statements which are true for your chosen number. (Just list the statement letters.)

- Then swap lists and see who can find the other's number first.

    (When you get too good at this version, you might like to start by choosing any integer less than 50, or less than 100.)

4 Can you find two numbers less than or equal to 30 which fit exactly the same statements? (Make sure you find **all** the statements that fit each of your two numbers.)

5 Find the number between 1 and 50 which has the largest number of statements that fit it. What do you think makes this number special?

6 **a** Which numbers less than 30 do **not** satisfy statement **X**?

    b Which numbers less than 50 satisfy both statements **N** and **Z**?

7 Any number which satisfies both statements **A** and **H** also satisfies statement **K**. Suppose a number satisfies both statements **B** and **D**; which other statement must it satisfy? Why must this other statement be satisfied?

8 **a** List all the statements that are satisfied by the number 1. How many of these statements (or rather, how **few** of them) do you need to define the number 1 **uniquely**?

   b How many of the statements **A–Z** do you need to uniquely define the number 2?

   c Repeat for each of the numbers 3–12.

EXTRA 2

Smarties

Every unopened Smarties tube is less than half full! Is this true?

Solution: page 78

8   *Paper folding : Exploring holes*

This activity focuses on:

- thinking about properties of polygons and folding;
- trying to predict in advance what will happen in geometrical experiments;
- making sense of unexpected outcomes.

Organisation/Before you start:

- This work is best done by two or three pupils working together.
- Each group will need small amounts of scrap paper.

Try to complete the whole section using at most 12 pieces of A6 size paper.

Fold a piece of paper in half like this:

Fold it once more, either straight: or at an angle:

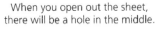

Then choose one 'cut line' and cut off the folded corner:

When you open out the sheet, there will be a hole in the middle.

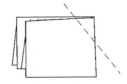

1   You are allowed to fold and cut exactly as above: two folds, then one cut.

    a Can you make a square-shaped hole?
       (Don't use trial and error. Think carefully how to fold and cut in order to get the hole you want.)

    b Can you make a rhombus-shaped hole that is not a square?

    c What other quadrilateral-shaped holes is it possible to make?

    d Can you make a triangle-shaped hole?

    e What kind of triangle-shaped holes is it possible to make?

    f Can you make a hole with more than four sides?

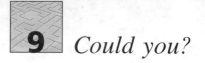

9 _Could you?_

This activity focuses on:

- deciding what calculation to do to answer a numerical question;
- practising estimation;
- looking up numerical facts;
- working with units in context.

Organisation/Before you start:

- You will need access to suitable reference books.
- If you are working with a partner, pick some questions to work on individually, then explain your methods and conclusions to each other.

The person answering these questions is assumed to be an ordinary human being with unlimited ability to organise, but no supernatural powers.

When answering these 'Could you ?' questions, you should make sensible **estimates**, and then **calculate** efficiently. Do not just guess! The important part of this exercise is how you simplify the required calculation.

Could you ...

1 ...fit the population of London into 1000 double-decker buses?

2 ...run a kilometre in one minute?

3 ...write your name and address on a piece of paper with an area of 0.01 square metres?

4 ...make a single pile containing one million sheets of paper in your school hall?

5 ...eat exactly one tonne of food in a year without becoming either very thin or very fat?

6 ...reach as high as the Sears Tower in Chicago if all the pupils in your school stood on each others' shoulders?

7 ...lift a 30 litre bottle of water?

8 ...walk one hundred thousand miles in your lifetime?

9 ...fit one thousand drink cans in one cubic metre?

10 ...swim in 100 litres of water?

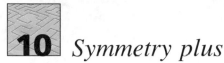

10 *Symmetry plus*

This activity focuses on:

- reviewing and extending work on symmetry;
- solving problems involving reflection and rotation, and the idea of similarity.

Organisation/Before you start:

- You will need some squared paper to make and cut out the shapes in question **5**. You will also need some A4 scrap paper.

1 **a** Only certain positions of the hands on an analogue clock are possible for a 'real time'. If the minute hand points exactly at the 12, what are the possible positions for the hour hand?

   b An analogue clock shows the time to be 9:30. If you reflect the hands in the line joining the 12 o'clock mark to the 6 o'clock mark, what time will the clock show? How much **later** is the new time than the old time?

   c Given any position of the hands which represents a real time, reflect the hands in the line joining the 12 o'clock mark to the 6 o'clock mark to get a new position for the hands. What is the maximum difference you can get between the old time and the new time? What is the minimum difference you can get between the old time and the new time?

   d Given any position of the hands which represents a real time, reflect the hands in the line joining the 12 o'clock mark to the 6 o'clock mark. Does the new position of the hands **always** represent a real time?

2 **a i** Fold an A4 sheet in half along its longest line of symmetry. The new smaller rectangle is **not** similar to the one you started with. Why not?

   ii Fold an A4 sheet in half along its shortest line of symmetry. The new smaller rectangle **is** similar to the one you started with. Why is this true?

   b Which other familiar shape has the property that you can fold it exactly in half to produce a shape which is similar to the shape you started with?

3 Start with a square and fold it in half twice to produce another square. That is, after two folds you get a shape which is similar to the shape you started with.

Find another familiar shape that you can fold in half twice and end up with a shape which is similar to the shape you started with.

4 I have folded a shape in half twice and have ended up with a parallelogram which is neither a rectangle, nor a square. What shape could I have started with? How many different possibilities are there for this starting shape?

5 Copy these three shapes onto squared paper and cut them out.

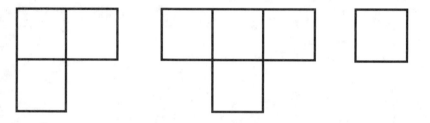

Put all three pieces together, so that they meet edge-to-edge, and so that the complete shape has either reflection or rotation symmetry. How many different symmetrical shapes is it possible to make? (There are a lot more than you might think, so you will have to find a systematic way of making a complete list.)

EXTRA 3

Clock hands

A clock says a quarter past two. How often during the next 24 hours will the two hands lie exactly on top of each other?

Solution: page 78

11 BIGGER *or smaller*

This activity focuses on:

- understanding the relationship between similar triangles;
- asking related 'What if ?' questions and following them through.

Organisation/Before you start:

- You will need A4 paper, a ruler, and a protractor/angle measurer.

Mark four points at random on a piece of paper.

Label three of the points A, B, C and join them up to form a triangle *ABC*.
Label the fourth point D.

Measure the distance from D to A, and mark A' beyond A, with D, A, A' in a straight line and $DA' = 2 \times DA$.

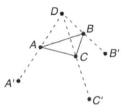

Repeat this for B and C to obtain the points B', C' with D, B, B' and D, C, C' in a straight line and

$$DB' = 2 \times DB, \quad DC' = 2 \times DC$$

Join up the three new points A', B', C' to form a new triangle A'B'C'.

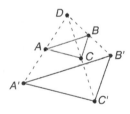

1 Write down as many relationships as you can

   a between the lengths of the sides of triangle *ABC* and the sides of triangle *A'B'C'*

   b between the angles of triangle *ABC* and the angles of triangle *A'B'C'*

   c between the shapes of the two triangles *ABC* and *A'B'C'*

   d between the perimeters of the triangles *ABC* and *A'B'C'*

   e between the areas of the triangles *ABC* and *A'B'C'*.

2 What would the answers to question **1a–e** be if the fourth point D was chosen **inside** the triangle *ABC*?

3 What if, instead of choosing A' 'beyond' the point A with $DA' = 2 \times DA$, you go **backwards** (with $DA' = 2 \times DA$)?

4 What if the new point A' was marked so that D, A', A lie in a straight line and $DA = 2 \times DA'$ (and similarly $DB = 2 \times DB'$, $DC = 2 \times DC'$)?

5 What if you wanted to produce a triangle $A'B'C'$ with each side 'three times as long', or 'one third as long' as the corresponding side of triangle ABC?

EXTRA 4

Shadows

Is it possible to make an object which:

- casts a circular shadow when light shines on it from the top, and

- casts a square shadow when light shines on it from the side, and

- casts a triangular shadow when light shines on it from the front?

Solution: page 78

12 *What do you think it's all about?*

This activity focuses on:

- making sense of a passage of text;
- thinking about surveys and questionnaires.

Read through this passage. Then answer the questions.

Surveys are used to find out what people do or think. They have been   **1**
carried out for many years. One of the first was to find out about the
poor in London. It was carried out by Charles Booth in the 1890s.
Governments used opinion polls during World War II.

*Large scale surveys are expensive, so they should be as **reliable** as*   **5**
possible. Before writing the questions you need to decide who to ask
and the type of survey to use. There are three main types:

(a) *face-to-face;*

(b) *telephone;*

(c) *postal questionnaire, which **subjects** complete for themselves and*   **10**
then post back in a pre-paid envelope.

Whichever method is chosen the questionnaire should be easy to
*understand and to complete. The **raw results** need to be simple to*
*make **processing** easy. For this reason **tick boxes** are very popular.*
Some people prefer four responses, such as   **15**

    ☐ *strongly agree*   ☐ *agree*   ☐ *disagree*   ☐ *strongly disagree*

For some survey questions a ☐ *'don't know' box may be useful.*

Even with this extra box strange survey results can sometimes be
obtained. About 50 years ago an American survey asked the question
'What do you think about the new metallic metals law?'. There was no   **20**
such law! Yet over 75% of the subjects ticked an opinion box, and less
*than 25% ticked 'don't know'! Questionnaires are usually **pre-tested***
with a small number of subjects to try to eliminate such problems.

1   Explain in your own words the meaning of the terms in **bold**.

2   What might a government want to find out from an opinion poll
     in wartime? (line 4)

3   What are the advantages and disadvantages of having four
     response boxes rather than five? (lines 14–17)

4   What are the advantages and disadvantages of using postal
     questionnaires? (lines 10–11)

13 *Proof*

This activity focuses on:

- developing an awareness of what 'proof' means;
- writing proofs of conjectured general patterns using simple algebra.

Organisation/Before you start:

- You need a copy of a 10 by 10 one hundred square.
- You need to be able to multiply out brackets before doing question **5**. So be prepared to stop after completing questions 1–4.
- Be prepared to discuss things with your teacher to resolve difficulties, and to sort out the difference between pattern-spotting and proof.

1	2	3	4	5	6	7	8	9	10
11	12	13	14	15	16	17	18	19	20
21	22	23	24	25	26	27	28	29	30
31	32	33	34	35	36	37	38	39	40
.	.	.	.	.	.	.	.	.	.
.	.	.	.	.	.	.	.	.	.
81	82	83	84	85	86	87	88	89	90
91	92	93	94	95	96	97	98	99	100

1 Pick any 3 by 3 square from the 10 by 10 one hundred number square.

24	25	26
34	35	36
44	45	46

a In your chosen 3 by 3 square:
Add the top left number and the bottom right number.
Then add the top right number and the bottom left number.
What do you notice about the two answers?

Suppose you chose a different 3 by 3 square from the 100 number square. If you did exactly the same (add top left and bottom right, then add top right and bottom left) would you always get equal answers? Explain.

b Start again. Pick any 3 by 3 square.
Add the middle number in the top row and the middle number in the bottom row.
Then add the middle number in the left-hand column and the middle number in the right-hand column.
What do you notice about your two answers?

Suppose you chose a different 3 by 3 square. If you did exactly the same, would you still get equal answers? Explain.

c Now add the numbers at the four corners of your 3 by 3 square.
Then add the four numbers in the middle of the four edges.
Try to explain why the two answers will always be equal no matter which 3 by 3 square you start with.

One way to check whether something is true for **every** 3 by 3 square in the 100 number square is to check each of the $8 \times 8 = 64$ possible 3 by 3 squares in turn! This is very slow, and you would be very likely to make a mistake along the way. And after all that work you could still not be sure what happens for 3 by 3 squares which lie outside the one hundred number square – such as the one shown here.

94	95	96
104	105	106
114	115	116

In mathematics, if you want to prove something general, you need **algebra**.

2   In question **1a** you want to know whether something is true for **every** 3 by 3 square.
    You want to calculate without using particular numbers.
    So write the letter n in the top left-hand corner of an unknown 3 by 3 square.

    a If the number in the top left corner is n, what is the next number in the middle of the top row?
    And what is the number next to that in the top right-hand corner?

    b Now write the numbers in each of the other six squares in terms of n.

    c Use this to **prove** that what you noticed in question **1a** was not an accident by carrying out these three steps:

    • Write down the algebraic expression involving n for the sum
    $$\begin{pmatrix} \text{number in the} \\ \text{top left corner} \end{pmatrix} + \begin{pmatrix} \text{number in the} \\ \text{bottom right corner} \end{pmatrix}.$$

    • Then write down the expression for the sum
    $$\begin{pmatrix} \text{number in the} \\ \text{bottom left corner} \end{pmatrix} + \begin{pmatrix} \text{number in the} \\ \text{top right corner} \end{pmatrix}.$$

    • Simplify each expression to show that they are both equal to $2n + 22$.

    This **proves** that the two sums are always equal, no matter what the value of n is.

    Now read through the **Comments and solutions** for question **2c**.

3   Use the approach of question **2c** to **prove** the results you found in question **1b** and in question **1c**.

4   Now start again and consider an unknown 3 by 3 square.
    Write the letter m in the centre square.
    Find the other numbers in the 3 by 3 square in terms of m.
    Hence solve questions **2c** and **3** in a different way.

5 Take any 3 by 3 number square from the 100 square.

- Multiply the numbers in the top left and bottom right corners.

- Then multiply the numbers in the bottom left and top right corners.

- Prove that the two answers will always differ by 40.

6 An **even** integer is one which is exactly divisible by 2.
So even integers are precisely those integers that can be written as '2 × an integer', that is in the form

$$2n, \text{where } n \text{ is some other integer.}$$

An **odd** integer is one that leaves remainder 1 when you divide by 2.
So the odd integers are precisely those that can be written in the form

$$2k + 1, \text{where } k \text{ is some other integer.}$$

Use these ideas to show that the sum of an even integer and an odd integer is always odd.

7 a It is a fact that **the sum of two even integers is always even**. But the 'proof' shown here is wrong. What is wrong with it?

> **Proof**: Let the two even integers be x and y.
> Then x is even, so $x = 2n$ for some integer n.
> And y is even, so $y = 2n$ for some integer n.
> $\therefore x + y = 2n + 2n = 4n$ which is even. **QED**

b Write out a correct proof which shows that the sum of two even integers is always even.

c What can you say about the sum of two odd integers?
Prove your claim.

8 Draw a cross ✕. Choose two numbers and put them at the top.

Add the two numbers and put the sum at the centre of the cross. Add the numbers along each sloping line and put the answers at the bottom end of that line.
Add the two bottom numbers; add the two top numbers. How are these two answers related? Can you **prove** it?

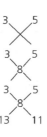

14 *Plane shapes*

This activity focuses on:

• planning an experiment and carrying it through;
• collecting suitable data, analysing it and sorting out the results.

Organisation/Before you start:

• You will need plain paper, a ruler, scissors, and glue.
• This work is best done by two or three pupils working together.

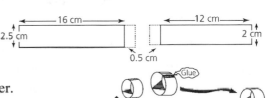

1   Make a full-size copy of this rectangle.

Fold it in half lengthwise; then fold it in half again to give three clear creases.

2   Fold the rectangle and glue the fourth flap to make a triangular girder.

3   **a** Make full-size copies of these rectangles.

b Make each rectangle into a cylindrical loop.

Stick one loop at each end of the triangular girder.

The task:
To make and test a paper plane; to modify your design; to carry out controlled experiments in order to find a design with the best performance.

The question:
What design, and what method of launching works **best?**
(For example: Does it fly best with the smaller loop at the front, or at the back? Is it better to launch the plane the right way up, or upside down?)

Testing:
Decide what to test, and how to test it fairly.
(For example: How exactly will you decide what is best? Should you measure length of flight? Or the smoothness/stability/straightness of flight? How much data should you collect? How accurate must your measurements be? How will you display your data? What kind of averages are you going to use? How sure can you be that the results from one design really show that it is better, or worse, than another?)

15 *More likely than probable*

This activity focuses on:

- identifying and clarifying some simple ideas about probability;
- resolving some common misunderstandings.

Organisation/Before you start:

- This work **must** be done by two or more pupils working together.

1 Select any one of the statements **A–Q** on p.29.

   a The statement you have chosen may be either correct or incorrect. Discuss the statement amongst yourselves: **you have to decide whether it is right or wrong**.
It may help to ask:

- Is there something you already know that might help you decide?

- Can you think of any examples that would help you to decide?

- Can you simplify the situation being described in some way that would help you to understand the question better?

- Is there an experiment you could carry out to help you test the statement?

- Can you draw some kind of diagram that might help?

   b Write down what you as a group decide about the statement you chose.

- If you think it is correct, then you should explain how you know.

- If you think it is incorrect, then you should explain what is wrong with it, and give an example to show how it goes wrong.

Make sure that what you write is not just your opinion. Explain why your decision is correct.

2 Now choose another of the statements **A–Q** and repeat question **1a** and **1b**. (You are not expected to assess all seventeen statements!)

A There are 13 people in a rugby league team, so it is certain that in any team there will always be two players who have their birthday in the same month.	B When two coins are tossed the probability of getting one Head and one Tail is $\frac{1}{2}$.
C In many board games you have to throw a 6 before you can start. They always ask for this because a 6 is the hardest number to get when you throw a dice.	D A fair way of choosing a number from 0–9 is to open a book at random and to take the units digit of the right-hand page number.
E When you throw a fair six-sided dice you are unlikely to get a prime number.	F If Notferry Athletic play Manchester United, Notferry can win, draw, or lose. So the probability that Notferry will win must be $\frac{1}{3}$.
G Bob says that he likes multiple-choice questions because he is bound to get half of them right just by guessing.	H When you throw a fair six-sided dice it is harder to get a 6 than it is to get a 3.
I I have just tossed a fair coin nine times and it showed a Head each time. If I now toss it a tenth time, it is more likely to show a Tail than a Head.	J A fair ten-sided dice, with faces numbered 1–10, is thrown twice. The number you get on the second throw will depend on which number turned up on the first throw.
K On the next School Sports Day it is more likely to rain than not to rain.	L I am more likely to win a raffle if I choose two raffle tickets from different places in the book, than if I choose two consecutive tickets.
M Whenever I drop a piece of buttered toast, it always lands butter-side-down.	N If I drop a drawing pin it can land either point up or point down. So the probability of landing 'point up' is $\frac{1}{2}$.
O If I blow on a dice before I throw it, I am more likely to get a 6.	P In the Lottery, the set of six numbers 2, 23, 25, 34, 40, 44 is more likely to come up than the set 1, 2, 3, 4, 5, 6.
Q In any rugby match there is a **better than even chance** that two of the players have the same birthday.	

More likely than probable

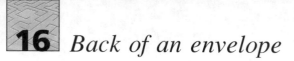

16 *Back of an envelope*

This activity focuses on:

- making sensible assumptions and estimates;
- obtaining good approximations by using rounding and simple arithmetic;
- handling large numbers and completing hard calculations with confidence.

Organisation/Before you start:

- This work is best done by two or more pupils working together.

If you combine simple approximations and elementary arithmetic you can find good estimates to answer all sorts of interesting questions. This kind of calculation is often called a 'back of the envelope' calculation because it is so simple it could be done on the back of an envelope.

The first question is an almost complete worked example to help you answer the other questions. All you have to do is to copy the statements and work out what goes in the spaces.

1 The table below shows the 1993 population of the UK split into different age groups. The numbers are given in millions.

under 16	16 to 39	40 to 64	65 to 79	80 and over	Total
12	20	17	7	2	58

a Roughly how many pupils in the 11–16 age range attend school in the UK?

Here is one way to answer this kind of question.

- According to the table, there are _____ million children under 16.

- Averaging this over 16 years, we get for each school year roughly $(12 \div 16)$ million children: $12 \div 16 =$ _____.

- 11–16 schooling takes _____ years.

- So there are about $5 \times 0.75 =$ _____ million pupils aged 11–16.

b About how many teachers are needed to teach UK pupils in the 11–16 age range?

Here is one way to answer this question.

- If you make allowance for non-teaching time, there are roughly 20 pupils for each teacher.

- So the number of teachers needed to teach 3.75 million pupils is roughly $3\,750\,000 \div 20 =$ _____.

c Check these estimates against official statistics.

Use these same ideas to answer questions **2–11**.

- Use sensible estimates and simple calculations. **Don't just guess**.

- Write down the assumptions you make when estimating, and the method you use when calculating.

2 Roughly how many secondary schools are there serving the 11–16 age group in the UK?

3 a Roughly how many primary school teachers must there be in the UK?

  b How many primary schools do you think there are in the UK?

4 a Roughly how many sides of paper do you use during a school day?

  b Roughly how many sides of paper do you use each week?

  c Roughly how many sides of paper does this work out at per year?

  d **Assume** that an exercise book contains around 80 sides of paper. About how many exercise books do you get through in a year?

  e About how many **maths** exercise books would you expect pupils in the 11–16 age group in your school to use each year?

5 a Use your answer to question **4d** to estimate the number of exercise books used by 11–16 pupils in UK secondary schools in a year.

  b It takes about 15 trees to make a tonne of paper. About how many trees are cut down each year just for exercise books for UK secondary schools?

6 Assume there are about 23 million households in the UK. For each of the following items, write down the additional assumptions you need to make in order to estimate how many of each item there are in the UK (you may find it easier to work in millions):

a television sets

b telephones

c milk bottles

d pairs of jeans

e pairs of scissors

f pairs of high-heeled shoes

7 Roughly how many burgers are eaten by pupils from your school in a year?

8 How many cans of soft drink are drunk in your town or city each week?

9 How much water is flushed away in toilets, or from baths, in the UK in a single day?

10 How many chips are eaten each day at lunchtime in UK schools?

11 On average each household in the UK is sent something like two 'junk mail' letters each week. How many trees does all this mail use up in a year?

Now make up some similar questions of your own. Then do the relevant calculations to find approximate answers.

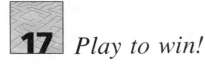

17 *Play to win!*

This activity focuses on:

- following rules;
- understanding the idea of a strategy;
- analysing simple games and devising winning strategies.

Organisation/Before you start:

- This work should be done by two pupils working together.

In some games there is a **strategy** for winning. In other games, though you cannot make sure of winning, you can at least make sure you don't lose!

In each of the games described here you have to try to find a strategy either **i** to win every time, or **ii** to avoid losing.

The most effective strategies are those that are easy to remember and to use. Sometimes it helps to draw a diagram to analyse the game.

A: Vingt et un

- Two players take turns to hold up any number of fingers from 1 to 5.
- They keep a running total of the number of fingers shown so far.
- The winner is the first player to get the running total to 21.

Play the game several times. Take turns to go first.
Can you find a strategy so that the first player always wins?

B: Prime or out

This game is just like **Vingt et un** except that the aim of the game this time is for each player to ensure that the running total is always a **prime number**. The first player to produce a running total which is **not** a prime number loses.

a What numbers can the first player choose?

b Is there always a winner? Or can the game go on for ever?

c Can the first player somehow make sure of winning every time?
Or can the second player force the first player to lose?

C: Dots and Crosses

Play this next game with an opponent several times:

- Put seven dots in a line.

- Take turns to cross out one dot that has not been crossed out already.

- The first player to complete a line of three crosses in a row wins (the three crosses do not all have to be made by one player).

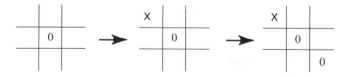

a Is it better to go first, or to go second?

b What if you start with a different number of dots?

D: Noughts and Crosses

According to a book on **Noughts and Crosses** you can always avoid losing if you:

- first grab the centre square

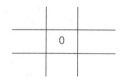

- then respond to each of your opponent's moves by choosing the 'opposite' square (that is, the square obtained by a half turn about the centre).

a Test the strategy. Does it work?

b Is it possible for either player to do better than just avoid losing?

18 *Computing*

This activity focuses on:

- learning about some of the people and the ideas involved in the early history of computing;
- developing research skills.

Organisation/Before you start:

- You will need access to a library, and will have to do some research to find the information you need.

1   This is a simplified image of a section of computer tape from the 1960s. The dots indicate small holes in the tape.

In those days, this kind of tape with patterns of holes was used to feed information and instructions into a computer.

There is a simple code which allows certain arrangements of holes to stand for certain letters.

a Find out how this code worked, and use it to decode the message on the piece of tape shown above. (If you get stuck, look up the **Comments and solutions**.)

b Make up a message of your own using this code, and challenge a friend to decode it.

2   **a** Find out all you can about **binary numbers**.

b Make up some sums using binary numbers. Work out how to do column addition and subtraction of binary numbers, and also long multiplication.

3   **a** Early computers used **binary code**. Why was this?

b Nowadays computers use **hexadecimal** arithmetic. What is this? How is it related to binary? Why was the change made?

4   **a** What methods are used today to input information or instructions into a computer?

b Find out some other ways which used to be used to input information into a computer. Why are these methods not used any more?

5 Find out what you can about the contribution made by one of these people to the way modern computers work:

Charles Babbage;   Alan Turing;   John von Neumann.

6 In a computing context, what are **bits** and **bytes**?

7 **a** Does the prefix **kilo** (as in kilobyte) have the same meaning in a computing context as it does in the word kilogram?

b In a computing context you will also find the letters M and G used as prefixes. What are these letters short for? What numbers do they represent?

8 Explain how it can be that **Oct 31** is the same as **Dec 25**.

EXTRA 5

The portrait

A man was looking at a portrait when someone asked 'Whose picture is that?' The man replied: 'Brothers and sisters have I none, but that man's father was my father's son.' Whose portrait was it?

Solution: page 78

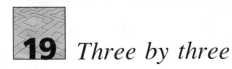

19 Three by three

This activity focuses on:

- developing and improving investigative skills;
- learning to look for ways of extending a given investigation.

All the activities in this section refer to the 3 by 3 grid on the right. You have to:

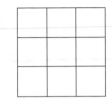

- choose one of the questions **A–J**;

- work through your chosen problem;

- develop and extend the idea as far as you can by asking and solving related 'What if . . . ' questions of your own.

A How many squares can you find?	**B** How many rectangles can you find (excluding squares)?

C By colouring in whole squares, how many different fractions of the whole grid can you show?	**D** How many different patterns can you make by colouring in whole squares
	
	    a using just one colour?
	
	    b using two colours?
	
	    c using three colours?

E Start with the 1×1 square in the centre. Round it is a border of eight 1×1 squares making the 3×3 grid.
 	
    a Now add a second border of 1×1 squares all round the 3×3 grid. How many squares would you need for the second border?
 	
    b How many 1×1 squares would you need to make the 100th border?
 	
    c How many 1×1 squares would there be altogether after the 100th border had been added?
	F **a** Suppose you can only move along grid lines, and that you are only allowed to move **downwards**, or to the **right**. How many different routes are there from the top left-hand corner to the bottom right-hand corner?
	
	    b How many routes are there from the top left corner to the bottom right if you are allowed to move in any direction provided that you never go over the same line twice?

G Enter the numbers 1–9, one number per square (in any order). Then add each pair of horizontally adjacent numbers. Finally add these six totals together to obtain a Grand Total T.
How should you position 1–9 so as to get the largest possible T?

H How much of the 3×3 grid can you draw while sticking to the grid lines,
and without taking your pencil off the paper,
and without going over any line twice?

I Can you enter the numbers 1–9 (in any order, one number per square) so that the resulting arrangement becomes a correct column addition sum, with two 3-digit numbers adding to give the total in the bottom row?

J Can you enter the numbers 1–9 in the grid so that no two adjacent numbers (that is, in squares that share an edge vertically or horizontally) have the same sum?

EXTRA 6

Units of . . .

Here is a list of some old-fashioned units.

a acre	**b** bushel	**c** carat	**d** furlong
e fluid ounce	**f** hundredweight	**g** gill	**h** grain
i league	**j** knot	**k** stone	**l** peck

Find out as much as you can about each one – including their origin, what they measured, the corresponding modern units which have replaced them, and details of how to convert measurements from these old-fashioned units to the corresponding modern units.

1 **a** Find a fraction which is greater than $\frac{1}{4}$ and less than $\frac{3}{10}$.

   b I am thinking of two fractions – but I won't tell you what they are. Can you tell me how to find a fraction which lies between my two fractions?

2 How many different sizes of squares can be made by choosing four dots in a 4 by 4 square array and then joining them up?

3 A cuboid has faces with areas $24\,cm^2$, $32\,cm^2$ and $48\,cm^2$. What is its volume?

4 How many triangles are there hidden in the pentagon on the right?

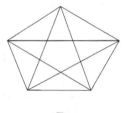

5 My watch is so accurate that it gains just 1 second per day. If it is correct now, how long will it be before it is correct again?

6 **a** On this 5 by 5 array of dots, can you find two rectangles which have equal areas, but unequal perimeters?

   b How many such pairs of rectangles are there in the 5 by 5 array?

7 How many 3 cm by 5 cm rectangles can be cut:

   a from a 6 cm by 10 cm rectangle?

   b from a 8 cm by 10 cm rectangle?

   c from a 7 cm by 11 cm rectangle?

   d from a 7 cm by 29 cm rectangle?

   e from a 17 cm by 22 cm rectangle?

1 *BIG numbers*

In these solutions we write ' $\approx$ ' to mean 'is approximately equal to'.

1 It is too fiddly to worry about leap years, so we take

$$1 \text{ year} \approx 365 \text{ days}$$

$$\therefore \quad 1 \text{ million hours} \approx \frac{1\,000\,000}{24 \times 365} \text{ years}$$

$$\approx 114.16 \text{ years}.$$

So it is very unusual for a human being to live that long.

If you simplify the calculation, you could even do it in your head!
24 is slightly less than 25, and 365 is slightly less than 400.

$$\therefore \quad 24 \times 365 \approx 25 \times 400$$

$$= 10\,000$$

$$\therefore \quad \frac{1\,000\,000}{24 \times 365} \text{ is slightly greater than } \frac{1\,000\,000}{10\,000} = 100.$$

2 The number of seconds in a day is

$$60 \times 60 \times 24 = 86\,400$$

$$\therefore \quad 1 \text{ million seconds} = \frac{1\,000\,000}{86\,400} \text{ days}$$

$$\approx 11\tfrac{1}{2} \text{ days}.$$

So after a million seconds, you would have had no birthdays at all – unless you count your zeroth 'birth' day.

You could also do this in your head:

$6 \times 6 = 9 \times 4$, and 24 is slightly less than 25.

$$\therefore \quad 60 \times 60 \times 24 = 100 \times 9 \times 4 \times 24$$

$$\approx 100 \times 9 \times 4 \times 25$$

$$= 900 \times 100 = 90\,000$$

$$\therefore \quad 1 \text{ million seconds is slightly greater than } \frac{1\,000\,000}{90\,000} = \frac{100}{9} \text{ days}.$$

COMMENTS & SOLUTIONS

maths CHALLENGE 1

40

3 A billion seconds is almost exactly $\dfrac{1\,000\,000\,000}{60 \times 60 \times 24 \times 365}$ years

This is about 31.7 years.

So if Christ was born at the beginning of 1 AD,

then a billion seconds later would be late in 32 AD.

4 A billion minutes is almost exactly $\dfrac{1\,000\,000\,000}{60 \times 24 \times 365}$ years.

This is slightly more than 1902.5 years.

So the answer to the question would be:

- 97 AD or 98 AD if you were answering during the year 2000, or

- 99 AD or 100 AD if you were answering during 2002.

5 It all depends how long it takes to say each number.

To keep things simple, suppose it takes 1 second to say each number.

Then it takes a million seconds to count to a million: that is, just over eleven and a half days (using the answer to question **2**).

In fact, it takes less than 1 second to say 9 and more to say '47 329'. Suppose you allow half a second **for each digit**: then '47 329' uses up $2\frac{1}{2}$ seconds. How would you then calculate a good estimate for the time needed to count to a million?

6 A pound coin has diameter roughly 22 mm, or $\frac{14}{16}$ inch.

$\therefore$ A million coins will reach about $\dfrac{22 \times 1\,000\,000}{1000}$ metres

$= 22$ km.

$\therefore$ Five million pound coins will reach about 110 km, or 69 miles.

This is roughly equal to the distance from Leeds to Nottingham.

7 The distance from Land's End to John O'Groats is about 850 miles by road, or 1350 km.

From question **6**, you know that a million coins will stretch about 22 km.

$\dfrac{1350}{22} \approx 61.4$.

$\therefore$ The number of coins needed is about 61.4 million.

8 A line of one million pound coins is 22 km long, or about 13.8 miles. So it would take roughly 4 hours and 36 minutes to walk past them all.

9 A pile of exactly ten one pound coins is about 33 mm tall.
∴ A pile of one million pound coins would be 3.3 km high.
This is much taller than the Eiffel Tower, which is only 300 m tall.

10 Ten one pound coins weigh 90 grams, or $3\frac{1}{4}$ ounces.

∴ A kilogram of pound coins contains $\dfrac{10 \times 1000}{90}$ coins.

∴ 1 kg of one pound coins is worth about £110.

11 A telephone box is just over 2 metres tall.
A pile of exactly ten one pound coins is about 33 mm tall.
$33 \times 60 = 1980$
1980 mm = 198 cm, which is almost 2 metres.
∴ A pile 2 metres high contains just over 600 pound coins.

A telephone box is about 75 cm wide and 70 cm deep.
A pound coin has diameter roughly 22 mm.

$\dfrac{750}{22}$ is just over 34, and $\dfrac{700}{22}$ is just over 31.

So it is possible to fit in about 34 piles one way and 31 piles the other.

This gives a grand total of just over $600 \times 34 \times 31 = 632\,400$ coins.

(If you arrange the piles to form an equilateral triangular array, you could fit in more, but not quite a million.)

2 *Pieces of pie*

1 **a** The 9s round the edge represent ninths.
The 8s represent eighths. And so on.

  b First make sure the pie is in the centre of the plate.
Then use the 7s (= sevenths), and cut from each 7 mark to the centre.

2 Two fractions are equivalent if they use the same mark.

  a Start at the top and move clockwise round the plate.

The second 6 coincides with the first 3: this says that 2 sixths is the same as 1 third.

(6 sixths make a whole; so 3 sixths make a half, and 2 sixths make a third.)

$\therefore \frac{2}{6}$ and $\frac{1}{3}$ are equivalent.

b, c 9 ninths make a whole; so 3 ninths is the same as 1 third, and 6 ninths is the same as 2 thirds.

$\therefore \frac{3}{9}$ and $\frac{1}{3}$ are equivalent,

$\therefore 2 \times \frac{3}{9} = \frac{6}{9}$ and $2 \times \frac{1}{3} = \frac{2}{3}$ are equivalent.

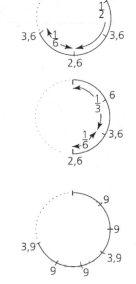

3 a $\frac{1}{2} + \frac{1}{6} = \frac{3}{6} + \frac{1}{6} = \frac{4}{6} = \frac{2}{3}$

b $\frac{1}{4} + \frac{3}{8} = \frac{5}{8}$

c $\frac{2}{3} + \frac{1}{6} = \frac{5}{6}$

d $\frac{1}{2} - \frac{1}{6} = \frac{1}{3}$

e $\frac{3}{4} - \frac{1}{8} = \frac{5}{8}$

f $\frac{2}{3} - \frac{1}{9} = \frac{5}{9}$

4 a (How many sixths are there in one half?) $\frac{1}{2} \div \frac{1}{6} = 3$

b (How many ninths are there in two-thirds?) $\frac{2}{3} \div \frac{1}{9} = 6$

c (How many sevenths are there in one whole?) $1 \div \frac{1}{7} = 7$

3 *Equals*

1 'Witte' means the same as 'Wit', 'know-how', or 'intelligence'. A 'whetstone' is a smooth stone used to sharpen knives. By calling his book on arithmetic and algebra *The Whetstone of Witte*, Recorde is suggesting that these subjects, properly learned and used, are a way of sharpening your intelligence!

2 He wrote (starting at the end of line 3): 'And to a voide the tediouse repetition of these woordes: 'is equalle to': I will sette as I doe often in woorke use a paire of paralleles, or Gemowe lines [meaning 'twin' lines] of one lengthe, thus:====, bicause noe .2. thynges, can be moare equalle.'

3 $=$: is equal to $\approx$: is approximately equal to
$\leq$: is less than or equal to $\neq$: is not equal to
$\geq$: is greater than or equal to $\equiv$: is identically equal (or
$\simeq$: is approximately equal to   equivalent) to

5 Here are some definitions you may come across:

- the same in size, amount, or value (Oxford paperback)

- (in an equation) whatever is on the left equals whatever is on the right; whatever comes before the $(=)$ is worth the same as whatever comes after it (Collins Maths)

- identical in quantity; of the same value (Chambers)

- related by a reflexive, symmetrical and transitive relationship; broadly alike or in agreement in a specific sense with regard to specific properties (Reader's Digest).

6 **a** $\frac{4}{6}$ *is equivalent to* (or *is the same as*) $\frac{2}{3}$.

   b 4×5 *gives the same answer as* (or *is identically equal to*) 5×4.

7 A cup of coffee *costs* 45p. (A cup of coffee is never equal to 45p.)

8 1 cm *represents* 10 m. (1 cm is never equal to 10 m.)

9 π is *approximately equal to* 3.14. (π is not exactly equal to 3.14.)

10 Henry VIII *married* Anne Boleyn. [Family trees often write '$A = B$' to mean 'A married B'; Henry VIII and Anne Boleyn are in fact different people.]

11 $\bowtie\!$ *represents* (or *stands for*) 1000 fish.

12 $2.4807 \approx 2.5$, or $2.4807 = 2.5$ *to 1 decimal place*. (2.4807 is not actually equal to 2.5).

13 $4 + 5$ is definitely *not* equal to 9×3. A correct calculation might be laid out as below on the left, or on the right.

$$4 + 5 = 9 \qquad\qquad (4 + 5) \times 3 - 7 = 9 \times 3 - 7$$
$$9 \times 3 = 27 \qquad\qquad\qquad\qquad\qquad = 27 - 7$$
$$27 - 7 = 20 \qquad\qquad\qquad\qquad\qquad = 20$$
$$\therefore \quad (4 + 5) \times 3 - 7 = 20$$

14   = is being used in two different senses.

The = in each equation is correct.

The = at the start of each line is being *wrongly* used as a shorthand for 'which means'. This is sloppy: one should write '∴'.

A correct calculation should be laid out something like this:

$$3x + 4 = 19$$
$$\therefore \quad 3x = 19 - 4$$
$$\therefore \quad 3x = 15$$
$$\therefore \quad \ x = 5.$$

 4   *True, False and 'Iffy' number statements*

It is important to learn to make precise, mathematically correct statements.

To help show what is meant by a 'mathematically correct' statement, these answers are very full – even though this may make them a bit hard to read.

Definitely true: D, E, H, J, N ⎫ If you disagree with any of these, think again.

Definitely false: G, Q ⎬ If you remain unsure, ask your teacher.

'Iffy': A, B, C, F, I, K, L, M, O, P, R

Each **'Iffy'** statement has been made more precise by adding some underlined words to the original statement.

There are many different ways of making an **'Iffy'** statement precise. We give the most obvious correct version.

If your answer is not the same as the answer given here, you will have to decide whether your answer is correct.

A   The product of two <u>whole</u> numbers is a whole number,
or
The product of two <u>integers</u> is an <u>integer</u>.

The word **number** is too vague, since it is used to refer to all sorts of numbers – positive, negative, zero, fractions, decimals, etc. The original statement **A** is false because the word number includes fractions and decimals.

When we mean a (positive or negative or zero) whole number, it is better to use the word **integer**.

B Adding zero to the end of <u>an integer</u> multiplies it by 10.

C The square root of a number x is smaller than x <u>if and only if $x > 1$.</u>

The expression 'if and only if' indicates that we have combined the two simple statements

(i) **if** $x > 1$, **then** $\sqrt{x} < x$; and (ii) **if** $\sqrt{x} < x$, **then** $x > 1$

into one, more compact, statement.

F When you square any number <u>other than zero</u>, the answer is positive.

Positive means **greater than zero**; hence $0^2 = 0$ is not positive.

I <u>Every</u> prime number <u>greater than 2 is</u> odd.

K The sum of two <u>positive</u> numbers is greater than their difference.

For the sum of two numbers to be greater than their difference, the two numbers **have to be positive**.

Proof: Let the two numbers be x and y, where $x > y$.

Suppose that $x + y > x - y$.

$$\therefore y > -y \quad \text{(adding } -x \text{ to both sides)}$$

$$\therefore 2y > 0 \quad \text{(adding } y \text{ to both sides)}$$

$$\therefore y > 0 \quad \text{(multiplying both sides by } \tfrac{1}{2}).$$

Since $x > y$, it follows that x must also be greater than 0. **QED**.

L If the only factors of a perfect square are the square itself, its square root, and 1, <u>then the original number must have been the square of a prime number.</u>

M If the three integers are
<u>(a) 1, 2, 3, or (b) ‾1, ‾2, ‾3, or (c) ‾k, 0, k (where k is any integer), then</u> the product of three whole numbers is the same as their sum.

This may seem a bit weak. More interesting is the fact that the **converse** of this statement is also true:

'<u>If</u> the product of three whole numbers is the same as their sum, then the three integers must be
(a) 1, 2, 3, or
(b) ‾1, ‾2, ‾3, or
(c) ‾k, 0, k (where k is any integer)'.

O Dividing a number x by a number $y < 1$ gives a number <u>larger than x if and only if either</u>
<u>(a) x and y are both positive, or (b) x and y are both negative.</u>

P When you multiply two numbers <u>greater than 1</u>, the answer is bigger than either of them.

A common mistake is to remember the half truth: 'multiplication makes things bigger'.

R The cube <u>x^3</u> of a number <u>x</u> is bigger than its square x^2 if and only if $x > 1$.

Squares are always greater than or equal to 0: $\therefore x^2 \geq 0$.

If $x^3 > x^2$, then $x \neq 0$; $\therefore x^2 > 0$;
so we can divide both sides of $x^3 > x^2$ by x^2 to get $x > 1$.

If $x > 1$, then $x \neq 0$; $\therefore x^2 > 0$;
so we can multiply both sides of $x > 1$ by x^2 to get $x^3 > x^2$.

$\therefore x^3 > x^2$ if and only if $x > 1$.

5 Möbius bands

1 When you glue the two ends of the strip together after giving one end a half twist, the surface you get is not quite what you might think.

It **looks** as though it has **two** edges; however, if you look more carefully you will find that it has only **one** edge which winds round the band twice before joining up on itself.

When you cut all the way round along the dotted centre line you first expect the Möbius band to fall into two pieces – with the point marked *A* and point marked *B* in different pieces.
On second thoughts you should realise that point *A* will be part of the same piece as the edge *a*; and the point *B* will be part of the same piece as edge *b*.
But the Möbius band only has one edge!
So edge *a* and edge *b* are connected.
Hence, even after cutting, the side of the band which contains the point *A* is still connected to the side of the band which contains the point *B*!

The spatial relationships explored in the other questions are sometimes just as surprising as question **1**. However, once you have understood the comments on question **1** you should be in a position to think through what happens in the other questions and to make sense of what you find.

The three-dimensional version of a Möbius band is called a **Klein bottle** – named after the nineteenth century German mathematician Felix Klein (1849–1925). This is a 'closed surface' (that is, a surface without any edge) which has only one side.

There is more about Möbius bands and Klein bottles in these books.

Martin Gardner, *Mathematical puzzles and diversions*, Pelican 1965 (Chapter 7 Curious topological models)

Martin Gardner, *More perplexing puzzles and tantalizing teasers*, Dover 1988 (Problem 21 of the second half)

David Wells, *The Penguin dictionary of curious and interesting geometry*, Penguin 1991 (pp. 152–153)

6 *Fractions with, and without, a calculator*

2   **Method 1** Given two fractions a, b.

- Work out $a - b$ (or $b - a$).

- See whether $a - b$ is positive or negative:
  - (i) if $a - b > 0$, then $a > b$;
  - (ii) if $a - b < 0$, then $b > a$.

3   **Method 2** Given two fractions such as $\frac{13}{18}$ and $\frac{11}{15}$.

- Choose an integer N that is divisible by both the denominators 18 and 15 – say $N = 180$.

- Then $\frac{13}{18}$ of N is an integer (130), and $\frac{11}{15}$ of N is an integer (132). The larger of these two integers tells you which fraction is largest.

4   **Method 3** Given two fractions such as $\frac{13}{18}$ and $\frac{11}{15}$.

- Produce a list of fractions equivalent to $\frac{13}{18}$

$$\frac{13}{18} = \frac{26}{36} = \frac{39}{54} = \frac{52}{72} = \frac{65}{90} = \frac{78}{108} = \cdots,$$

and a list of fractions equivalent to $\frac{11}{15}$.

$$\frac{11}{15} = \frac{22}{30} = \frac{33}{45} = \frac{44}{60} = \frac{55}{75} = \frac{66}{90} = \cdots.$$

- Find fractions in each list with the same denominator $(\frac{13}{18} = \frac{65}{90}, \frac{11}{15} = \frac{66}{90})$.

- Then compare the fractions $\frac{65}{90}$ and $\frac{66}{90}$ with the same denominator to see which is largest.

5   **Method 4** Given two fractions such as $\frac{13}{18}$ and $\frac{11}{15}$.

- Change each fraction to an **approximate** decimal (by dividing the numerator by the denominator (or by using the F>D key if you have one):

$$\frac{13}{18} = 0.722\,222\,22\ldots, \frac{11}{15} = 0.733\,333\,33\ldots.$$

- Then compare the two decimal approximations to see which is the largest.

6   The following comments may help:

Method 1 To subtract one fraction from another, replace the two fractions by equivalent fractions with a **common denominator.**

You can then do the subtraction easily by subtracting the **numerators.**

Method 2 It is easier to work with $\frac{11}{15}$ of an integer A if the integer A is a multiple of 15 (since the answer is then an integer). It is easier to work with $\frac{13}{18}$ of an integer B if the integer B is a multiple of 18 (since the answer is then an integer).
To compare the two fractions $\frac{11}{15}$ and $\frac{13}{18}$ we want the two integers A and B to be equal: $A = B$.
So you want to find a single integer which is a **common multiple** of 15 and of 18.
Use the smallest common multiple of 15 and 18 that you can find. (The smallest possible common multiple is called the **least common multiple** or LCM of 15 and 18.)

Method 3 There is no need to produce a long list of equivalent fractions. Instead you can make use of the idea in Method 2, and choose a common multiple of 15 and 18.
Then write down one fraction equivalent to $\frac{11}{15}$ and one fraction equivalent to $\frac{13}{18}$ – each having the same denominator.

Method 4 The question asks you to make sure you understand how to carry out each method without using a calculator.
So you need to know how to work out $11 \div 15$ and $13 \div 18$ using *long division*.
You only need to continue each division far enough to allow you to compare the two decimal answers.
(In some cases you may only need one or two decimal places).

8 a Write the two fractions $\frac{2}{3}$ and $\frac{4}{5}$ using a common denominator:

$$\frac{2}{3} = \frac{2 \times 5}{3 \times 5} \text{ and } \frac{4}{5} = \frac{4 \times 3}{5 \times 3}.$$

It is then clear that the second fraction is bigger because $2 \times 5 < 4 \times 3$.

b Write the two fractions $\dfrac{11}{15} = \dfrac{11 \times 6}{15 \times 6}$ and $\dfrac{13}{18} = \dfrac{13 \times 5}{18 \times 5}$

using a common denominator.
It is then clear that the first fraction is biggest because $11 \times 6 > 13 \times 5$.

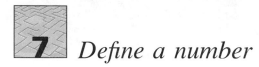

7 Define a number

1 a **B, L, S, X, Z**.

2 a 6, 10, 14, 15, 21, 22, 26.

b **B, G, H, J, L, P, R, W, X**.

c 12, 16, 20, 24, 28 have $2^2 = 4$ as a factor;
18, 27 have $3^2 = 9$ as a factor.

d 23, 29.

e **E** (1, 8, 27); **I** (3, 13, 23); **V** (9, 18, 27).
(Note: **T** fails because 7, 14, 21, 28 are all multiples of 7.)

3 a There is only one 'even prime number': 2.

b If a number is 'a square which is also divisible by 5' (**D** and
L), then it must be equal to the square of some multiple of 5.
If the number is also odd (**B**), then it must be equal to the
square of an *odd* multiple of 5.
If the number is also less than one thousand, then it must equal

$$5^2 = 25, \text{ or } 15^2 = 225, \text{ or } 25^2 = 625.$$

4 No. The statements have been deliberately chosen so that each
number between 1 and 30 is uniquely defined by the statements
which it satisfies. For example,

- 1 is the only positive integer satisfying **D**, **E** and **F**.

- 2 is the only positive integer satisfying both **A** and **C**.

- 3 is the only positive integer satisfying both **C** and **H**.

- 4 is the only positive integer satisfying **A**, **D** and **F**.

5 36 satisfies **A, D, G, H, K, P, Q, S, U, V, X, Z**. Two things about
36 that help it to fit so many of the statements are:

- 36 is a square;

- 36 has lots of factors (nine in fact).

6 a The only numbers less than 30 which do not satisfy statement
X are 1, 2, 3, 11, 17, 23, 27, 29.

b The only numbers less than 50 that satisfy both **N** and **Z** are 28
and 46.

7 Any integer which satisfies **B** and **D** must satisfy **O** (and this is the only other statement which *must* be satisfied).

Proof: The square of an even integer is always even.
So, if an integer N is odd (**B**) and a square (**D**), it must be the square of a smaller odd integer. This smaller odd integer can be written in the form $2n + 1$, for some integer n.

$$\therefore N = (2n + 1)^2$$
$$= (2n + 1) \times (2n + 1)$$
$$= 4n^2 + 4n + 1$$
$$= 4(n^2 + n) + 1.$$

Hence dividing N by 4 leaves remainder 1. **QED**

8 Here are the answers for 1, 2, 3, 4; the rest are left to you.

a 1 satisfies **B, D, E, F, O, P, R, W, Y;**
1 is the only positive integer satisfying **D, E** and **F**.

b 2 satisfies **A, C, F, R, W, Y, Z;**
2 is the only positive integer satisfying **A** and **C**.

c 3 satisfies **B, C, F, H, I, P, R, W, Y;**
3 is the only positive integer satisfying **C** and **H**.

4 satisfies **A, D, F, R, W, X, Y, Z;**
4 is the only positive integer satisfying **A, D** and **F**.

These questions are meant to help you think clearly and logically. You should tackle them systematically, but using ordinary thinking.
However, you could do all the thinking **first** – by making a table in which you take one statement at a time and check each integer in turn to see whether it satisfies that statement:
1 does not satisfy statement **A**, so put a cross;
2 does satisfy statement **A**, so leave blank; and so on.
You could then use your table to answer each question.
This would be very systematic (and may be a useful way to check your answers – especially in question **4**). But it would also remove the challenge 'to think on your feet'.

	1	2	3	4	5	6	7
A	×		×				
B		×		×			
C	×			×			
D		×	×				

8 *Paper folding: Exploring holes*

1 **a** Make two straight folds at right angles. Then cut at 45°.

b Make two straight folds at right angles. Then cut at an angle other than 45°.

c It may help to start by thinking about some specific quadrilaterals.

- Can you make make a kite-shaped hole that is not a rhombus?

- Can you make a parallelogram that is not a rhombus?

- Can you make a rectangle that is not a square?

This exploratory approach should convince you that it is possible to make lots of **kites** (make the second fold at lots of different angles – not just a right angle – before you cut). Other shapes seem impossible!

Think about the way you fold and cut.

Whenever folding and cutting produces a quadrilateral, pairs of adjacent sides of the quadrilateral are always folded and cut **together**.

These adjacent sides **must have the same length**, so the resulting shape has to be a **kite**.

If both pairs of equal sides have the same length as each other, the kite is a rhombus.

If adjacent sides are at right angles to each other, you get a square.

d As in part **c** it may help to start by thinking about some specific triangles.

- Can you make a triangle that is not isosceles?

- Can you make an acute-angled triangle?

- Can you make a right-angled triangle?

- Can you make an equilateral triangle?

To make a triangle-shaped hole, make the second fold at an angle.

Then cut **perpendicular to** part of the first fold.

e The triangles you get are always isosceles. Can you see why?

If you choose the second fold carefully, you can make the isosceles triangle acute-angled.

To produce a right-angled triangle, make the second fold at 45°.

To produce an equilateral triangle, make the second fold at 60°.

f If you cut as instructed, you cannot get a hole with more than four sides.

With two folds there are just four layers of paper, each of which is cut in a straight line.

These four cut layers produce four straight sides which are the edges of a polygon with at most four sides.

It is impossible to create **more than** four sides (though if you fold at an angle and then cut perpendicular to one of the fold lines – as in question **1d** – two of these straight cuts may join up to make a single side, so producing a triangle).

Extension

2 **a** Suppose you allow yourself three folds to produce a corner, and one cut across the corner. What shape holes can you make? (Predict before you cut!)

b What if you use just two folds, but allow two cuts?

9 *Could you?*

1 The population of London is about 7 million.
So there would have to be about 7000 people in each bus; this is impossible.

2 To run a kilometre in 60 seconds you would have to run each 100 m in 6 seconds.
The best sprinters take more than 9 seconds to run 100 metres, so this is impossible.

3 $1\,\text{m}^2 = 10\,000\,\text{cm}^2$

$\therefore 0.01\,\text{m}^2 = 100\,\text{cm}^2$

So $0.01\,\text{m}^2$ is the size of a $10\,\text{cm} \times 10\,\text{cm}$ piece of paper. You should find it quite easy to write your name and address on a piece of paper this size.

4 Your school hall could be as much as $10\,\text{m}$ high; it is certainly less than $20\,\text{m}$ high.

One hundred sheets of paper make a pile about $1\,\text{cm}$ high. So a pile of one million pieces of paper would be $10\,000\,\text{cm}$, or $100\,\text{m}$, high.

5 1 tonne $= 1000\,\text{kg}$. 1 year $= 365$ days.

So to eat 1 tonne in a year, you would have to average $\frac{1000}{365}\,\text{kg}$, that is nearly $3\,\text{kg}$, each day.

This amount would probably make you fatter!

6 The Sears Tower is $443\,\text{m}$ tall.

Each pupil would contribute about $1.5\,\text{m}$ (from the soles of their feet to their shoulders).

$\therefore 443\,\text{m}$ would need around $\frac{443}{1.5} \approx 300$ pupils.

7 One litre of water weighs one kilogram.

So a 30 litre bottle of water would weigh more than $30\,\text{kg}$. You might just manage to lift this, but you couldn't carry it very far.

8 Suppose you live for 80 years. We ignore the first five years (since you don't walk much during this period).

75 years $\approx 75 \times 365$ days

$= 27\,375$ days.

To walk a total of $100\,000$ miles, you would have to average $\frac{100\,000}{27\,375}$ miles each day, that is a little over $3\frac{1}{2}$ miles per day. Do you average that much **each day**?

9 A drink can has diameter $6.5\,\text{cm}$, so you could fit at least 15 in a row of length one metre.

$\therefore$ You could fit at least $15^2 = 225$ in a square metre.

Each can is just less than $11.5\,\text{cm}$ tall, so you could pile eight on top of each other without exceeding one metre in height.

$\therefore$ You could easily fit $225 \times 8 = 1800$ cans into one cubic metre.

10 To swim you would need a tank at least 1 metre wide and about 2 metres long.

$$1\,\text{litre} = 1000\,\text{cm}^3$$

$$\therefore\ 100\,\text{litres} = 100\,000\,\text{cm}^3.$$

So you could fill a 1 m by 2 m rectangular pool only to a depth of $\frac{100\,000}{100 \times 200} = 5$ cm. This would not be deep enough to swim in.

10 *Symmetry plus*

1 **a** The hour hand must point **exactly** at one of the hour marks: any other position does not represent a 'real time'.

b It would be 2:30, which is 5 hours **later**.

c The maximum possible difference is 6 hours (starting at 3:00 or 9:00).
The minimum possible difference is zero (starting at 6:00 or 12:00).

d If you start with a real time, the new position of the hands after reflection **is** always a real time. This is not obvious.

Proof: The minute hand moves $6°$ per minute, while the hour hand moves $0.5°$.
If the hour hand is h degrees past an hour mark, the minute hand must have moved $12h°$ beyond the 12 o'clock mark.
So if the starting time was M minutes **after** H o'clock, the reflected time will be M minutes **before** $(12 - H)$ o'clock.

QED

For example, 9:30, or 30 minutes after 9, becomes 30 minutes **before** $(12 - 9)$ o'clock, that is, 2:30.

2 **a** Suppose an A4 sheet is an a by b rectangle (with $a > b$).

i The folded rectangle has the same length a and half the width $\frac{b}{2}$.

The ratio   length : width   has changed

from (old rectangle) $a : b$

to (new rectangle) $a : \left(\dfrac{b}{2}\right) = 2a : b.$

So the rectangles are not similar.

ii In an A4 sheet, the width is greater than half the length.

So the folded rectangle has length b and width $\dfrac{a}{2}$.

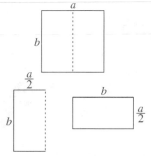

The ratio   length : width   changes

from (old rectangle) $a : b$,

to (new rectangle) $b : \dfrac{a}{2} = 2b : a.$

The two rectangles are similar if these two ratios are equal.

The dimensions of A4 paper are supposed to be chosen so that

$$\left(\frac{a}{b}\right)^2 = 2$$

$$\therefore \frac{a}{b} = \frac{2b}{a}$$

$$\therefore a : b = b : \left(\frac{a}{2}\right).$$

(This is only approximately true in practice. The dimensions of A4 paper are 297 mm by 210 mm, so $a/b = 297/210$. Work out $(297/210)^2$ to see how close the ratio $297/210$ is to $\sqrt{2}$.)

b Every right-angled isosceles triangle has the property. The line bisecting the right angle is a line of symmetry of the triangle. If you fold along this line, you produce another isosceles right-angled triangle.

3 Any rectangle has the same property.

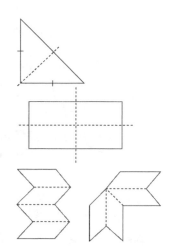

4 There are just two possible starting shapes: neither is obvious, but the first one shown here is easier to find than the second! If you interchange the long and the short sides of the parallelogram, you get two other possible shapes; but they are not really *different*.

5 There are seventeen – many more than you might think. Some shapes can be made in several different ways, though we give only one way of making each shape. The squares of the T-shape are marked as 1s, the squares of the L-shape are marked as 2s, and the single square as 3.

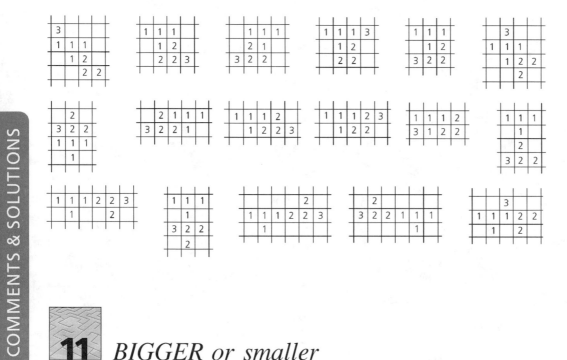

11 *BIGGER or smaller*

1 **a** Each side of the new triangle $A'B'C'$ is twice as long as the corresponding side of the original triangle ABC:

$$A'B' = 2 \times AB; \quad B'C' = 2 \times BC; \quad C'A' = 2 \times CA.$$

b Each angle of the new triangle $A'B'C'$ is the same as the corresponding angle of the original triangle ABC:

$$\angle A'B'C' = \angle ABC; \quad \angle B'C'A' = \angle BCA; \quad \angle C'A'B' = \angle CAB.$$

c The two triangles have the same shape, but different sizes. Triangle $A'B'C'$ is an **enlargement** of triangle ABC with **scale factor** 2.

d The perimeter of the new triangle $A'B'C'$ is exactly twice the perimeter of the original triangle ABC.

e The area of the new triangle $A'B'C'$ is exactly four times the area of the original triangle ABC.

2 The answers to question **1a–e** would be exactly the same. The only difference is that the new triangle $A'B'C'$ would now **contain** the original triangle ABC.

3 Once again, the answers would be exactly the same. The only difference is that, if the ordering of the vertices A, B, C of triangle ABC is clockwise, then the ordering of the vertices A', B', C' of triangle $A'B'C'$ would be anticlockwise: triangle $A'B'C'$ is an **enlargement** of triangle ABC with **scale factor** $^-2$.

4 The new triangle $A'B'C'$ is now **smaller** than the original triangle, with each side parallel to and exactly half as long as the corresponding side of the triangle ABC. Triangle $A'B'C'$ is an **enlargement** of triangle ABC with **scale factor** $\frac{1}{2}$.

5 To make each side of triangle $A'B'C'$ three times as long as the corresponding side of triangle ABC, you should mark the new points A', B', C' so that $DA' = 3 \times DA$; $DB' = 3 \times DB$; $DC' = 3 \times DC$.

To make each side of triangle $A'B'C'$ one third as long as the corresponding side of triangle ABC, you should mark the new points A', B', C' so that $DA = 3 \times DA'$; $DB = 3 \times DB'$; $DC = 3 \times DC'$.

12 *What do you think it's all about?*

1 **reliable:** they should find out *what people actually think* (rather than just what they say).

subjects: *the people* who are invited to respond to the questions.

raw results: *the actual responses* which subjects give to the questions before this information is processed.

processing: *performing calculations on the raw results* to produce data which is easier to understand. This includes collecting and grouping the responses, doing certain calculations with the numbers that emerge, and presenting the information graphically so that interesting trends can be clearly seen.

tick boxes: *options which only require* the subject to make *a simple choice* either to tick a box or to leave it, that is to decide Yes or No.

pre-test: *a trial run* of the questionnaire on a small group of subjects in order to iron out any 'wrinkles'.

3   It depends what the five boxes are.

- If don't know is used as a fifth box, it should be physically separated from the other four so that subjects see it as being a different kind of response.

- If you include a fifth box neutral between agree and disagree, then subjects tend to choose it to avoid making up their minds, so you don't get very useful information.

4   Postal questionnaires are cheap to carry out. They can also be targeted at certain groups. For example, you can often tell quite a lot about people's income or interests just from their postcode. On the other hand, you cannot control who will respond and who will not; so you can never be sure that those who respond are representative of the target group. For example, those who feel very strongly about something are more likely to respond; those who are very busy are unlikely to respond.

Extensions

5   See if you can find any postal questionnaires your family has been sent recently. Try to work out the purpose of some of the questions.

6   Design a questionnaire of your own to investigate the differences between boys' and girls'
    a attitudes to sport, or **b** attitudes to computers.

13 *Proof*

1   We give answers for the 3 by 3 square shown here.

   a First answer = 70; second answer = 70.
   The two answers are equal.

   b First answer = 70; second answer = 70.
   The two answers are equal. They are also equal to the answers in **a**.

24	25	26
34	35	36
44	45	46

c Both answers equal 140.
  To show these are not just accidents see the solution to question **2c**.

In working through this section you may have wondered whether there is a quick way to see that there are exactly $8 \times 8 = 64$ possible 3 by 3 squares in the one hundred number square.

In question **1** you could have chosen the
- the 3 by 3 square in the **top left** corner of the 100 number square; or
- the 3 by 3 square in the **bottom right** corner; or
- any 3 by 3 square in between.

Starting from the 3 by 3 square in the top left-hand corner,
- exactly eight 3 by 3 squares fit along the top row.

Starting from the same 3 by 3 square in the top left-hand corner,
- exactly eight 3 by 3 squares fit down the left-hand column.

The possible 3 by 3 squares form an 8 by 8 array, so there are $8 \times 8 = 64$ such 3 by 3 squares in the one hundred number square.

---

2 **a** The integers along the first row are: $n, n+1, n+2$.

  b The box immediately below n contains the integer $n+10$; so the integers along the middle row are: $n+10, n+11, n+12$.
  The number below $n+10$ is $n+20$; so the integers along the bottom row go: $n+20, n+21, n+22$.

  c Top left plus bottom right $= n + (n+22)$
  Bottom left plus top right $= (n+20) + (n+2)$.
  Both expressions are equal to $2n+22$.

3 Middle left plus middle right $= (n+10) + (n+12) = 2n+22$.
  Middle top plus middle bottom $= (n+1) + (n+21) = 2n+22$.

  Sum of integers at the four corners
  $= n + (n+2) + (n+20) + (n+22)$.
  Sum of integers in the middle of the four edges
  $= (n+1) + (n+10) + (n+12) + (n+21)$.
  Both sums are equal to $4n+44$.

4   Solution to question 1a
Top left $= m - 11$; bottom right $= m + 11$.
∴ top left plus bottom right $= (m - 11) + (m + 11) = 2m$.

The others are similar.

5   If we use the labelling from question **2**, we get

- (top left) $\times$ (bottom right) $= n \times (n + 22) = n^2 + 22n$
- (bottom left) $\times$ (top right) $= (n + 20) \times (n + 2)$
$$= n^2 + 20n + 2n + 40$$
$$= n^2 + 22n + 40.$$

These clearly differ by 40.

If we use the labelling from question **3**, we get

- (top left) $\times$ (bottom right) $= (m - 11) \times (m + 11)$
$$= m^2 - 11m + 11m - 121$$
$$= m^2 - 121$$
- (bottom left) $\times$ (top right) $= (m + 9) \times (m - 9)$
$$= m^2 + 9m - 9m - 81$$
$$= m^2 - 81.$$

These also differ by 40.

6   Proof: Let the even integer be x and the odd integer by y.
Then x is even, so $x = 2n$ for some integer n;
and y is odd, so $y = 2k + 1$ for some integer k.
∴ $x + y = 2n + (2k + 1)$
$$= 2n + 2k + 1$$
$$= 2(n + k) + 1,$$

which is odd since $2(n + k)$ is an even integer. **QED**

7   a The same symbol n has been used with two completely different meanings (n is used to stand for both one half of x **and** one half of y).
So the proof only makes sense if $x = y$.

b For a correct proof you must use different names for the different integers $\frac{x}{2}$ and $\frac{y}{2}$:
Proof: x is even, so $x = 2n$ for some integer n.
And y is even, so $y = 2k$ for some integer k.
∴ $x + y = 2n + 2k = 2(n + k)$,
which is even since $n + k$ is an integer. **QED**

c Claim: The sum of two odd integers is always even.

Proof: Let the two odd integers be x and y.

Then x is odd, so $x = 2n + 1$ for some integer n;

and y is odd, so $y = 2k + 1$ for some integer k.

$$\therefore x + y = (2n + 1) + (2k + 1)$$
$$= 2n + 1 + 2k + 1$$
$$= 2(n + k) + 2$$
$$= 2(n + k + 1),$$

which is even since $n + k + 1$ is an integer. **QED**

8 Let the two numbers at the top be a and b.

Then the number at the centre of the cross will be $a + b$.

The number at the bottom left will then be $b + (a + b) = a + 2b$,

and the number at the bottom right will be $a + (a + b) = 2a + b$.

So the sum of the two numbers at the bottom will be

$$(a + 2b) + (2a + b) = 3a + 3b = 3(a + b),$$

which is exactly 3 times the sum of the numbers at the top. **QED**

14 *Plane shapes*

There are no fixed answers in this section. The issues raised in these comments are meant to help you to clarify your thinking so you can make up your own mind.

A It is important in each test you use, that you should put each model through **exactly the same** trial. So you should try each time:

- to hold each plane in the same way;

- to use the same throwing action each time;

- to throw from the same height, in the same direction, with the same force, and at the same angle.

B Since you will not quite manage to make each trial exactly the same, it is important to repeat each throw a number of times, **and then to take the average of the results obtained**. But which 'average' should you use?

You want to avoid accidental bias, such as any slight change in the strength or angle of the throw. In many instances where a trial is repeated you might think of taking the **mean** of the results obtained. However, this does not allow for results where the thrower, or the wind, did something unusual. So it may be better to take the length of the **median** throw in each batch. Or to discard the longest and the shortest throws before taking the mean. Or to ...
You have to decide!

C In a truly scientific experiment, the thrower should probably be blindfolded, so that they do not give a little extra push when throwing their own favourite model!

D The experiments are best done indoors (to avoid the effects of wind). If you have to do them outdoors, then you will have to try to minimise, or to make allowance for, the effects of the wind.

E All throws should be made from a fixed start line – which could be marked simply by putting a book on the floor.
All distances should be measured as accurately as possible from this start line. If a long tape measure is not available, you could mark each metre along the intended flight path by placing a succession of separate markers on the floor.

F You will have to decide whether you measure the distance between the start line and the landing point, or whether you only measure the distance travelled in the **intended** direction of throw perpendicular to the start line.

Extensions

Once you have answered the initial questions using the official design, think how to vary the design. Make some planes which have a longer, or a shorter, body, or which have larger or smaller loops, than in the official design. Which flies best? How can you compare different designs? (Do they fly better with the smaller loop on top or underneath? Is it better with the two loops both on top, both underneath, or one on top and one underneath? What happens if the two loops have the same size? What if you put two loops side by side (**not** one on top of the other) at one or both ends?)

maths CHALLENGE 1

64

15 *More likely than probable*

A This is correct.
If players numbered 1–12 all have birthdays in different months, then between them they use up all twelve available months.
So either:

a two of players 1–12 have their birthdays in the same month, or

b player number 13 must share his birthday month with one of the players numbered 1–12.

B The statement is correct.
Think of the coins as having different colours: say one red and one blue.
Then there are four possible outcomes – each of which is equally likely:

(red) Head and (blue) Head; (red) Tail and (blue) Head;
(red) Head and (blue) Tail; (red) Tail and (blue) Tail.

Two of these four outcomes have 'one Head and one Tail', so the probability of getting one Head and one Tail is $\frac{2}{4}$, that is $\frac{1}{2}$.

C This is incorrect.
When throwing a fair dice, all six numbers are equally likely.
So the probability of getting any particular number is exactly $\frac{1}{6}$.

D This is incorrect.
For example, most books start on the right-hand page, so page 1 would then be on a right-hand page.
But this means that page 3, page 5, and so on will all be right-hand pages, so that you will never get an even units digit on the right-hand page!
(If the book happens to start on a left-hand page, the method will produce only even digits.)

E The statement is incorrect.
Three of the six numbers on a dice are prime numbers – namely 2, 3, 5.
So the probability of getting a prime number when you throw a fair dice is exactly $\frac{3}{6}$, that is $\frac{1}{2}$.

F Although there are only three possible outcomes, they are not all equally likely, so the probability is most unlikely to be $\frac{1}{3}$.

G This would only be correct if:

a each question offers exactly two choices, and

COMMENTS & SOLUTIONS

b Bob chooses entirely at random.

Most multiple-choice papers offer four or five choices, so Bob is likely to be very wrong.

H   The statement is incorrect.
When you throw a fair dice each of the six numbers is equally likely to appear, so both 3 and 6 have probability $\frac{1}{6}$.

I   The statement is incorrect.
The whole point of tossing a coin is that the coin does not know what happened last time it was tossed.
So the probability of getting a Tail is exactly the same each time, no matter how many Heads may have appeared in a row.

J   The statement is incorrect.
The whole point of a fair dice (whether it has six or ten sides) is that all outcomes are equally likely each time since the dice does not know what happened on previous throws.
So the probability of getting a particular score with a fair ten-sided dice is exactly $\frac{1}{10}$ each time you throw the dice, no matter what score appeared on the previous throw.

K   This statement is probably incorrect (though it often appears to be true!).

L   This is incorrect – provided the tickets in the drum are properly mixed up. However, choosing from different places in the book makes it more likely that you will have one ticket which is close to the winning ticket.

M   Our own private research shows that this is false – though (like **K**) this sometimes appears to be true!

N   For most ordinary drawing pins this is incorrect.
A drawing pin is slightly more likely to land point up than it is to land point down.
(See John Durran, *Statistics and probability*, Cambridge University Press 1970, pages 19–25.)

O   This is incorrect.
If you blow on a dice the probability of each outcome remains exactly $\frac{1}{6}$.

P This is incorrect.
Each set of six numbers has exactly the same probability of coming up. This probability is approximately $\frac{1}{14\,000\,000}$.

Q The statement is correct.
Whenever there are at least 23 people, there is a probability greater than $\frac{1}{2}$ that two of them share exactly the same birthday! (The proof of this fact is elementary, but is too complicated to go through here.)
Each rugby union team has 15 players; each rugby league team has 13 players. So a rugby match involves either 30 or 26 players. In a football match, there are just 22 players, so there is still almost an even chance that two players share the same birthday, but it is slightly less than $\frac{1}{2}$.

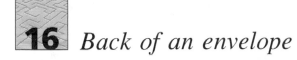

16 *Back of an envelope*

Estimates depend on the assumptions you make, and on the approximations you make when calculating. You are likely to have made slightly different assumptions from the ones used here, and you may also have made different approximations when calculating. **So your answers will be slightly different from those given here**. Do not worry about small differences; but it is worth going through the solutions here carefully to make sure you understand how any differences arise.

1 a About 3.75 million pupils.

b About 187 500 teachers.

2 Secondary schools vary in size.
Also some have pupils from 11–16, some from 11–18, some from 13–18 and so on. Suppose a typical secondary school has around 800–900 pupils in the 11–16 age range.
There are around 3.75 million pupils in the 11–16 age range (question **1a**).
∴ Number of schools with pupils in the 11–16 age range
$\approx \frac{3\,750\,000}{850} \approx 4500$.

3 a In some parts of the UK primary schooling lasts for 7 years. But for most children, primary schooling lasts 6 years. To get a rough estimate, assume that primary schooling last for 6 years.

There are about 0.75 (almost 0.8) million pupils in each school year.

∴ Number of primary school pupils $\approx 6 \times 0.8$ million; that is, just under 5 million.

Suppose there is about 1 teacher for every 25 pupils.

∴ Number of teachers in primary schools $\approx \frac{5\,000\,000}{25} = 200\,000$.

b Primary schools vary in size; the smallest have less than 50 pupils, but some have 400 or more pupils.

Assume that on average a primary school has 250 pupils. You already know that there are around 5 million primary pupils.

∴ Number of primary schools in the UK $\approx \frac{5\,000\,000}{250} = 20\,000$.

4 Your answers will depend on your own school. Remember that there are about 40 weeks in the school year.

After calculating your estimates, check your answers by asking your teacher.

5 **a** Multiply your answer to question **4d** by the approximate number of 11–16 pupils in the UK: you may take the estimate 3.75 million from question **1**, or approximate this to 4 million to make the arithmetic easier.

This assumes that your own personal use of paper is *on average* roughly the same as everyone else's – whereas in practice some people write much smaller than others, and some are very wasteful or messy!

You may be able to find someone who will tell you roughly how many exercise books are used by 11–16 pupils in your school each year (in all subjects): this would average out the differences between pupils.

You could then use this to estimate the number of exercise books used each year in a *typical* 11–16 school. To get a better answer to part **a**, multiply this number by the number of schools in the UK serving pupils in the 11–16 age range – which you know, from question **2**, is around 4500.

b An exercise book weighs about 100 g, so ten exercise books weigh 1 kg. A tonne is 1000 kg. So you should divide your estimate in part **a** for the total number of exercise books used each year by 10 000 to get the number of tonnes of paper used each year.

To get the number of trees cut down each year for this purpose, multiply the number of tonnes of paper needed each year by 15.

6 The following approximate figures may be useful.

a On average 1–2 TV sets per family.

b On average 1 and a bit telephones per family.

c On average each family may have 2 bottles of milk delivered each day – while a further 2–4 bottles are being cleaned and filled ready for the next day.

d Perhaps 2–4 pairs of jeans on average per family.

e Perhaps 3 pairs of scissors per family.

f On average perhaps 2–3 pairs of high-heeled shoes per family.

9 A single toilet flush uses about 10 litres of water. A bath uses about 200 litres of water. In a population of nearly 60 million, each person is likely to flush the toilet about four times a day; and perhaps one quarter to one half of the people have a bath each day. So ...

10 The total school population is around 12 million. There are about 20 chips in a portion. If half of all pupils eat school meals, and one half of these have chips, then ...

11 Suppose one piece of junk mail weighs around 50 g. So two letters per household per week gives around 100 g per household per week. There are just over 50 weeks in the year, so this gives 5 kg per household per year. Since there are around 23 million households, this gives 115 000 000 kg, or 115 000 tonnes – which is around $15 \times (115\,000)$ trees: that is, nearly one and three quarter million trees.

17 *Play to win!*

A: Vingt et un

If you leave a running total ≥ 16 you lose immediately.
This means that if you leave a running total between 10 and 14 you will lose three goes later (since your opponent will move the total on to $16 - 1 = 15$, and force you to leave a total ≥ 16).

This in turn means that if you leave a total between 4 and 8, you will lose in five goes' time (since your opponent will move the total on to $10 - 1 = 9$, and force you to leave a total between 10 and 14). Hence the first player has a simple strategy:

- Start by choosing $4 - 1 = 3$.

- Then, whenever the second player chooses 1, the first responds by choosing 5 $(1 + 5 = 6)$; whenever the second player chooses 2, the first responds by choosing 4 $(2 + 4 = 6)$; etc.

B: Prime or out

This is a bit harder than Vingt et un, but the idea is the same.

a The first player must put up either 2, 3 or 5 fingers.

b For there to be no winner, it would have to be possible **always** to get from one prime number total to the next by adding at most 5. However, there are gaps in the run of prime numbers – such as the jumps from 23 to 29, or from 31 to 37, or from 89 to 97 – which prevent this.

c The first gap of more than 5 is the jump from 23 to 29. So if a player can be sure of getting the total up to 23, then his/her opponent will lose. The first player can in fact make sure of getting the total to 23; but to do this only one of the three possible first moves (2, 3, or 5) works.

  i Which one is it?

  ii How many options does the second player then have?

  iii What must the first player choose next?

  iv How many options does the second player then have?

  v What must the first player choose next?

  vi How many options does the second player then have?

  vii What must the first player choose next?

C: Dots and Crosses

a It is best to go first. One simple and effective strategy for the first player is to choose the centre dot. Can you see why this is a good strategy? The second player then has two choices: either

- the second player decides **not** to choose one of the two end dots – in which case the first player wins immediately; or

- the second player chooses an end dot – in which case the first player has just one safe reply, and then wins on his or her next turn.

D: Noughts and Crosses

a The suggested strategy sounds promising – like the successful strategy in Dots and Crosses. But it doesn't work!
Suppose X makes the three moves shown, and 0 simply mirrors X's moves.
Then X gets a line of three one move before 0.

		X
0	X
		X

b Not if their opponent plays correctly.

If you want to consult books about strategies for games of this kind, it may help to know that the game we call Noughts and Crosses is also known as **Tic Tac Toe**.

Extension

Suppose you change the rules of Noughts and Crosses by deciding that the first player to produce a line of three noughts or three crosses **loses**. Is it possible for one player to avoid losing? Can one player be sure of winning?

 **18** *Computing*

Only three answers are given here since the whole point of this item is for you to go and find the information for yourself.

1 a Each **column** of holes, or dots, represents a letter.

- First, letters are encoded as numbers using
  A = 1, B = 2, C = 3, etc.

- Each of these numbers is then written in **binary** notation For example, W is first replaced by 23; then 23 is written as 10111.

- Finally this number is encoded as a single column of five 'digits' (where the two binary digits '1' and '0' are encoded by the presence or absence of a large hole in the tape) – with the top position being the units position, the next position being the '2s' position, the next being the '4s' position, and so on.

For example, W = 23, 23 = 10111 (in binary), so W is encoded as shown here.

$$\bullet$$
$$\bullet$$
$$\bullet$$

$$\bullet$$

You should now be able to decode the message.

3 b Hexadecimal arithmetic is arithmetic with numbers written in 'base 16'. Since $16 = 2^4$, a number expressed in binary form uses up four times as many characters as a number expressed in hexadecimal.
(Some modern calculators have the option of working in hexadecimal.)

7 Kilo stands for 10^3. M stands for **mega**, meaning million, or 10^6. G stands for **giga**, meaning billion, or 10^9.

8 It would be a pity to give away this little riddle. (It should be more than enough to hint that **Dec** is being used as shorthand for base_____.)

19 *Three by three*

A There are nine $(= 3^2)$ 1 by 1 squares;
four $(= 2^2)$ 2 by 2 squares;
and one $(= 1^2)$ 3 by 3 square.

So there are 14 $(= 3^2 + 2^2 + 1^2)$ squares altogether.

(What happens on a 4 by 4 grid? What about grids of other sizes?)

B There are six 2 by 1s, and six 1 by 2s;
three 3 by 1s, and three 1 by 3s;
two 3 by 2s, and two 2 by 3s.
So there are 22 altogether.

C How you tackle this will depend on whether you interpret the challenge as

- colour a single whole square
(either a 1 by 1 square – covering exactly $\frac{1}{9}$ of the grid, or a 2 by 2 square – covering exactly $\frac{4}{9}$ of the grid, or a 3 by 3 square – covering exactly $\frac{9}{9}$ of the grid); or

- colour as many 1 by 1 squares as you like
  (in which case you can clearly colour 0, 1, 2, 3, 4, 5, 6, 7, 8 or
  9, 1 by 1 squares, and so can cover any of the ten fractions
  $\frac{0}{9}, \frac{1}{9}, \frac{2}{9}, \frac{3}{9}, \frac{4}{9}, \frac{5}{9}, \frac{6}{9}, \frac{7}{9}, \frac{8}{9}, \frac{9}{9}$ of the grid).

D a Suppose just one colour is used, and that 'whole squares'
means 'whole 1 by 1 squares'.
Then each of the nine 1 by 1 squares is either coloured or not
coloured. So there are 2 choices for the first square (coloured
or not coloured); then 2 choices for the next square; and so on.
This means that there are
$2 \times 2 \times 2 \times 2 \times 2 \times 2 \times 2 \times 2 \times 2 = 512$ possible patterns
altogether.

b Suppose two colours (say red and blue) are used.
Then each of the nine 1 by 1 squares can be coloured red, or
coloured blue, or left uncoloured.
There are 3 choices for each square, and so
$3 \times 3 \times 3 \times 3 \times 3 \times 3 \times 3 \times 3 \times 3 = 3^9 = 19\,683$ possible patterns
altogether.

E a After adding the first border you have a 3 by 3 grid of 1 by 1
squares. The number of squares added is then $8 = 3^2 - 1^2$
(i.e. the difference between the number of 1 by 1 squares in the
final 3 by 3 grid and the number of 1 by 1 squares in the
original 1 by 1 grid).
After adding the second border you have a 5 by 5 grid of
1 by 1 squares. So the number of squares added is equal to
$5^2 - 3^2 = 16$ (i.e. the difference between the number of 1 by 1
squares in the final 5 by 5 grid and the number of 1 by 1
squares in the 3 by 3 grid).

b After adding the 100th border you have a 201 by 201 grid of
1 by 1 squares. So the number of squares added is equal to
$201^2 - 199^2 = (201 - 199) \times (201 + 199) = 2 \times 400 = 800$
(i.e. the difference between the number of 1 by 1 squares in the
final 201 by 201 grid and the number of 1 by 1 squares in the
previous 199 by 199 grid).

c After adding the 100th border there would be $201^2 = 40\,401$
1 by 1 squares.

F a Turn the 3 by 3 grid through $45°$ clockwise; the top left-hand
corner of the original grid is now at the very top.

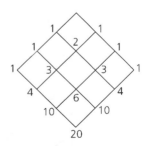

Any allowable route from the top to a given junction X must go via one of the two junctions A, B immediately above X. Hence every route to junction X goes either via A or via B; so the number of ways of getting to junction X is the sum of the numbers of ways of getting to the two junctions A and B.

This explains why the numbers which occur are like the numbers in **Pascal's triangle**.

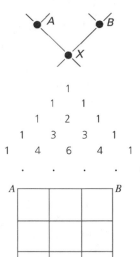

b If the only restriction is 'not to go over the same line twice', there is no nice way of counting. And there are far more routes than you would expect.

For example, in part **a** there is exactly one route from A to C which starts by going directly from A to B. But in part **b** there are **more than 50** routes from A to C which start by going direct from A to B!

G For the largest possible Grand Total the three largest numbers 7, 8, 9 must be in the middle column (so that each of them is counted twice).

Any such arrangement will have Grand Total equal to

$$1 + 2 + 3 + 4 + 5 + 6 + 2 \times (7 + 8 + 9) = 69.$$

H You are forced to miss three of the line segments. An example is shown on the right.
This is because you have to change 6 of the 8 junctions in a 3×3 grid where an **odd number** of lines meet. The difficulty is related to the famous **Bridges of Königsberg** problem which was solved by the Swiss mathematician Leonhard Euler around 1730.

I Here is one example.
Can you see how to change it to get another example?

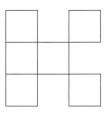

J Here is one example.

20 *Puzzles*

1 a $\frac{1}{4} = \frac{10}{40}$ and $\frac{3}{10} = \frac{12}{40}$. Hence $\frac{11}{40}$ is greater than $\frac{1}{4}$ and less than $\frac{3}{10}$.

    b In general, given any two numbers a and b, their **mean** $\dfrac{(a+b)}{2}$ lies exactly halfway between a and b.

2 There are *five* different sized squares: 1×1, 2×2, 3×3, and two 'sloping' squares as shown.

3 Suppose the cuboid has dimensions x cm by y cm by z cm. Then

$$xy = 24, \quad yz = 32, \quad zx = 48.$$

Multiplying all three equations together we get

$$xy \cdot yz \cdot zx = 24 \times 32 \times 48.$$

The left-hand product has two xs, two ys and two zs, so is equal to $(xyz)^2$. The right-hand side is equal to $(8 \times 3) \times (4 \times 4 \times 2) \times (8 \times 3 \times 2) = (8 \times 4 \times 3 \times 2)^2$.
$\therefore xyz = 8 \times 4 \times 3 \times 2 = 192$, so the cuboid has volume $192\,\text{cm}^3$.

4 The pentagon contains eleven separate 'pieces': **10** small triangles and **one** smaller pentagon in the centre.

Each of these ten small triangles forms a two-piece triangle together with the next small triangle clockwise – giving **10** two-piece triangles.

These small triangles fit together in threes at each vertex to make **5** three-piece triangles.

Two small triangles and the central pentagon fit together to make **5** different three-piece triangles.

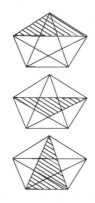

Finally four small triangles and the central pentagon fit together to make **5** five-piece triangles.

This gives **35** triangles altogether.

5 Between now and the next time my watch is exactly correct it must gain exactly 12 hours.
In 12 hours there are exactly $12 \times 60 \times 60$ seconds.
So I must wait exactly $12 \times 60 \times 60$ days: that is, I will have to wait **118 years and 130 days** (not counting leap years).

6 **a** The easiest pair to find is a 2×2 square and a 4×1 rectangle: both have area 4 square units, but the 2×2 square has perimeter 8 units and the 4×1 rectangle has perimeter 10 units.

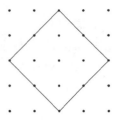

b There are five such pairs: 2×1 and $\sqrt{2} \times \sqrt{2}$; 2×2 and 4×1; 5×1 and $\sqrt{5} \times \sqrt{5}$; 4×2 and $\sqrt{8} \times \sqrt{8}$ (see figure); 5×2 and $\sqrt{10} \times \sqrt{10}$.

7 **a** You can fit exactly **4** copies of a 3×5 rectangle into a 6×10 rectangle.

b If you start with an 8×10 rectangle, the problem is more difficult.

- Notice first that since the 3×5 rectangle has area $15 \, \text{cm}^2$ and the 8×10 rectangle has area $80 \, \text{cm}^2$, you cannot possibly cut more than five copies of the small rectangle out of the big rectangle.

- However, it is not clear that you can actually get as many as five copies of the 3×5 rectangle (if you try to cut as in part **a**, then you will only get four copies together with a wasted L-shaped border).

- In fact, it is not hard to see that it **is possible to cut out five** copies of the small rectangle from the large one.

c If you start with a 7×11 rectangle, the problem is harder still.

- Notice first that since the 3×5 rectangle has area $15 \, \text{cm}^2$ and the 7×11 rectangle has area $77 \, \text{cm}^2$, you cannot possibly cut more than five copies of the small rectangle out of the big rectangle.

- However, it is not clear that you can actually get as many as five copies of the 3×5 rectangle: **getting four is easy**, but no matter which way you try, there never seems to be enough room to squeeze in the fifth.

- In fact, it is not hard to see that it is impossible. Notice there is no way of fitting 3s and 5s together to make 7. So in each of the eleven 7×1 strips in the 7×11 rectangle at least one unit square is always wasted – leading to a total waste of at least 11 unit squares – which leaves at most $66 \, \text{cm}^2$ available.

So **no matter how you cut, you cannot cut out five 3×5 rectangles**.

d Suppose you start with an 7×29 rectangle.

- Notice first that the 3×5 rectangle has area $15\,\text{cm}^2$ and the 7×29 rectangle has area $203\,\text{cm}^2$. So you cannot possibly cut more than 13 copies of the small rectangle out of the big rectangle.

- However, it seems impossible to get anything like 13 copies in practice. You can certainly fit ten 3×5 rectangles into a 6×25 rectangle in the obvious way – and since this leaves a 7×4 rectangle spare on one end, you can certainly cut out an eleventh small rectangle.
  So you **can cut out at least eleven 3×5 rectangles**.

- That this is in fact the maximum is shown by the argument we used in the solution to part **c**. As in part **c** at least one unit square is bound to be wasted in each of the 29 7×1 strips so the maximum available area is really only $174\,\text{cm}^2$. So **it is impossible to cut out as many as twelve 3×5 rectangles**.

e Cutting 3×5 rectangles from a 17×22 rectangle is even more interesting.

- If you look at areas, then $\dfrac{17 \times 22}{3 \times 5} < 25$, so you cannot possibly cut out more than 24 3×5 rectangles.

- But it is far from clear whether you can actually manage to find a way of cutting out 24 3×5 rectangles from a 17×22 rectangle.
  In fact **it is possible to cut out 24 3×5 rectangles**.

EXTRA 1

Washing up

In one week the number of pence you would earn is
$$1 + 2 + 2^2 + 2^3 + 2^4 + 2^5 + 2^6 = 2^7 - 1 = 127,$$
that is £1.27.

In two weeks you would earn
$$1 + 2 + 2^2 + 2^3 + 2^4 + \ldots + 2^{13} = 2^{14} - 1 \text{ pence},$$
that is, £163.83.

In three weeks you would earn $2^{21} - 1$ pence: that is, £20 971.51.
In four weeks you would earn $2^{28} - 1$ pence:
that is, £2 684 354.55.

EXTRA 2

Smarties

Yes it is true. One way to show this is to:

- take two unopened tubes of Smarties;

- open the two tubes and pour caster sugar carefully into each tube (shaking gently) until both tubes are full to the brim;

- then pour the contents of both tubes into an empty bowl, separate out the smarties and see if the remaining caster sugar will fit into one empty tube.

EXTRA 3

Clock hands

The hands cross at 12 o'clock; then again just after five past one; then again just after 10 past two; and so on. In fact the hands cross every $1\frac{1}{11}$ hours, so they cross 11 times every 12 hours – that is, 22 times in the next 24 hours.

EXTRA 4

Shadows

Here is such an object.

EXTRA 5

The portrait

The portrait is of the man's own son.

Glossary

- This list of symbols, and of common terms and their meanings, contains the information you need to make sense of the questions and the solutions.
- You may need to refer to a mathematics dictionary or a textbook for more detailed explanations.

Symbols

$a > b$	a is greater than b
$a \geq b$	a is greater than or equal to b
$a < b$	a is less than b
$a \leq b$	a is less than or equal to b
$a = b$	a is equal to b
$a \neq b$	a is not equal to b
$a \approx b$	a is approximately equal to b
$\sqrt{a}$	the square root of a
$a \equiv b$	a is equivalent to b (or a is identically equal to b)
$\therefore$	therefore
$\Rightarrow$	implies that

Terms

Acute An **acute** angle lies between $0°$ and $90°$. An acute-angled triangle is a triangle in which all three angles are acute.

Adjacent **Adjacent** means next to. Two sides of a triangle, or of a polygon, are adjacent if they meet at a vertex.

Analogue clock An **analogue clock** has a face with twelve hour marks equally spaced round the circumference. It has an hour hand and a minute hand which rotate about the centre of the clock face.

Average An **average** of a collection of numbers is a single number which manages to capture some important feature of the whole collection of numbers. The most commonly used average is the *mean;* other examples are the *median* and the *mode.*

Base We usually write numbers in **base** 10, using the digits 0, 1, 2, 3, 4, 5, 6, 7, 8, 9, with the value of each digit depending on its position: in the number 1234 (base 10), the 1 stands for 1 thousand ($1000 = 10^3$), the 2 stands for 2 hundreds ($100 = 10^2$), the 3 stands for 3 tens, and the 4 stands for 4 units.

Billion In the English-speaking world **billion** now means a thousand million, or 10^9. (It used to mean a million million, or 10^{12}, and it still has this meaning in some European languages, such as German.)

Binary	One can also write numbers in *base 2*, or **binary** – the only digits are then 0 and 1, and the binary number 10111 is equal to $1 \times 2^4 + 0 \times 2^3 + 1 \times 2^2 + 1 \times 2 + 1 = 23$.
Common	The integers 12 and 15 both have 3 as a factor: 3 is called a **common** factor of 12 and 15. The integer 60 is a multiple of 12; it is also a multiple of 15, so 60 is a *common multiple* of 12 and 15. Two adjacent sides of a triangle share a *common vertex*.
Consecutive	Two whole numbers are **consecutive** if they differ by 1 (e.g. 4 and 5). Two terms of a number sequence, or two events in a sequence of events, are consecutive if one comes immediately after the other.
Converse	Let N be an integer. Consider these two statements: (1) if N is the square of a prime number, then N has exactly three factors; (2) if N is the cube of a prime number, then N has exactly four factors. Both statements are true: if $N = p^2$ for some prime number p, then the only factors of $N = p^2$ are 1, p, and p^2; if $N = p^3$, then we get the extra factor p^3. The **converse** of statement (1) is: (1′) if N has exactly three factors, then N is the square of a prime number. This happens to be true. The **converse** of statement (2) is: (2′) if N has exactly four factors, then N is the cube of a prime number. This happens to be false (since 6 also has four factors).
Cube	The word **cube** refers to: (a) a solid shape with six square faces, or (b) a number like $27 = 3 \times 3 \times 3 = 3^3$, which is equal to the volume of a solid cube with integer side length 3. We say: 27 is the cube of 3.
Decimal place	In any decimal, the position immediately after the decimal point is called the first **decimal place** (that is, the tenths digit). The second position after the decimal point (that is, the hundredths digit) is the second **decimal place**.
Denominator	In the fraction $\frac{3}{4}$, the number on the bottom, 4, is the **denominator**. (The number on the top is the *numerator*.)
Difference	The **difference** of two numbers is found by subtracting the smaller number from the larger: the difference between $^-2$ and 4 is 6.
Digit	A **digit** is one of the numbers 0, 1, 2, 3, 4, 5, 6, 7, 8, 9. The number 1234 has four digits: 3 is the tens digit of 1234.
Divisible	63 is **divisible** by 9 (since $63 \div 9 = 7$ with no *remainder*). 63 is *not* divisible by 8 (since $63 \div 8 = 7$ remainder 7).

Enlargement	In triangle $A'B'C'$ each edge is parallel to, and twice as long as, the corresponding edge in triangle ABC. Triangle $A'B'C'$ is an **enlargement** of triangle ABC with *scale factor* 2.

Equation	An **equation** is a mathematical sentence in which two numbers or expressions are linked by an $=$ sign. The equation is correct only if the two numbers, or expressions, on either side of the $=$ sign have the same value.
Equivalent	Two different-looking fractions are **equivalent** if they represent the same number. For example, $\frac{2}{3}$ and $\frac{4}{6}$ are equivalent fractions.
Even chance	If an experiment has two possible outcomes, and each is equally likely, we say that there is an **even chance** of observing either outcome.
Expression	An **expression** is a collection of mathematical symbols (usually standing for known and unknown numbers, constants and variables), which are added or multiplied together, subtracted or divided. For example, πr^2 is an expression for the area of a circle of radius r; $2(l + w)$ is an expression for the perimeter of a rectangle of length l and width w.
Factor	9 is a **factor** of 63, because 63 can be written as '9 $\times$ an integer' ($63 = 9 \times 7$). 9 is not a factor of 64, because 64 cannot be written as '9 $\times$ an integer'.
Fair	A dice is **fair** if each of its faces is equally likely to show when the dice is rolled or thrown. A coin is fair if the two possible outcomes, Heads and Tails, are equally likely to occur when the coin is tossed.
Fraction	A **fraction** is the number obtained when one integer quantity (the dividend, or *numerator*) is divided by a non-zero integer quantity (the divisor, or *denominator*).
Glossary	A **glossary** is a list of specialist words, together with their meanings.
Integer	An **integer** is any whole number, whether positive, negative or zero; for example, $^{-}12$, 0, 5, 746.
Isosceles	A triangle ABC is **isosceles** with apex A if $AB = AC$; the side BC opposite the apex A is called the base of the isosceles triangle.

Kite	A **kite** is a quadrilateral $ABCD$ in which $AB = BC$ and $CD = DA$.
Least common multiple	The **least common multiple** (LCM) of two given integers is the smallest integer which is an exact multiple of the two given integers; for example, the LCM of 26 and 91 is 182.
Lowest common denominator	The **lowest common denominator** of two given fractions is the LCM of the two denominators. For example, the lowest common denominator of the fractions $\frac{5}{26}$ and $\frac{17}{91}$ is 182; both fractions can be written with this common denominator ($\frac{5}{26} = \frac{35}{182}$ and $\frac{17}{91} = \frac{34}{182}$), and this is the smallest such denominator.

GLOSSARY

Mean	The **mean** of the five numbers 7, 3, 2, 7, 11 is obtained by adding the five numbers $(7 + 3 + 2 + 7 + 11)$ and then dividing the answer, 30, by the number of numbers in the original list, 5. So the mean of the numbers 7, 3, 2, 7, 11 is 6.

Median	To find the **median** of the five numbers 7, 3, 2, 7, 11, put the numbers in order (2, 3, 7, 7, 11), then choose the middle number, 7. To find the median of the six numbers 8, 2, 1, 6, 29, 67, put the numbers in order (1, 2, 6, 8, 29, 67) and choose the mean of the two middle numbers, $(6 + 8)/2 = 7$.

Möbius band	A **Möbius band** or Möbius strip, is obtained by • starting with a narrow strip of paper whose ends are about to be glued (to make a short, fat cylinder), and then • twisting one of the two ends through $180°$ before glueing. The result is a surface with only one face and only one edge.

Mode	The **mode** of the five numbers 7, 3, 2, 7, 11 is the number which occurs most frequently: in this case 7.

Multiple	12 is a **multiple** of 3, since 3 is a *factor* of 12 ($12 = 3 \times 4$). $^-18$ is also a multiple of 3 (since $^-18 = 3 \times (^-6)$). 13 is not a multiple of 3.

Numerator	In the fraction $\frac{3}{4}$, the number on the top, 3, is the **numerator**. (The number on the bottom is the *denominator*.)

Parallel	Two (infinite) straight lines in the plane are **parallel** if they never meet.

Parallelogram	A **parallelogram** is a quadrilateral $ABCD$ in which the side AB is parallel to DC, and the side BC is parallel to AD.

Percentage	To refer to a specific fraction of a given quantity, such as 'one quarter of a rectangle', or 'two thirds of a class of 30 pupils', we can express this fraction as a **percentage**. To calculate the percentage corresponding to a given fraction, re-write the fraction with denominator 100; the numerator gives the percentage. For example $\frac{1}{4} = \frac{25}{100}$, so 'one quarter of a rectangle' is the same as '25% of the rectangle'.

Perfect square	An integer is called a *square*, or a **perfect square** if it is equal to the product of some integer by itself; for example, $0 = 0 \times 0$, $1 = 1 \times 1$, $4 = 2 \times 2$. So an integer is a perfect square if it is equal to the area of a geometrical square with integer length sides.

Perpendicular	Two straight lines are **perpendicular** if they are at right angles to each other.

Pictogram	A **pictogram** is a frequency diagram which uses a simple picture to stand for a fixed number of items.

GLOSSARY

Plane	A **plane** is a flat surface – that is, a surface in which the straight line joining any two of its points lies wholly in the surface.
Polygon	A **polygon** with three vertices (and three sides) is a triangle; a polygon with four vertices is a quadrilateral; a polygon with five vertices is called a pentagon, one with six vertices is a hexagon, one with eight vertices is an octagon, and one with ten vertices is a decagon.

A polygon *ABCDE...M* is an ordered sequence of points *A, B, C, D, E, ..., M* (called *vertices*) and line segments *AB, BC, CD, DE, ..., LM, MA* (called edges) in the plane, such that any two consecutive edges, such as *AB* and *BC* (or *MA* and *AB*), have exactly one vertex in common, and two non-consecutive edges have no points in common at all.

Prime number	A **prime number** is a positive integer which has exactly two factors – namely itself and 1. 1 is not a prime number.
Probability	Given a range of possible outcomes for an event, the **probability** of a particular outcome is an exact measure, given by a number between 0 and 1, of how likely that outcome is. For example, when tossing a fair coin, there are just two possible outcomes: heads and tails. With a fair coin these two outcomes are equally likely, so each outcome has probability exactly $\frac{1}{2}$. An outcome which is completely certain has probability 1; an outcome which is impossible has probability 0.
Product	Given a collection of numbers, their **product** is the number obtained when you multiply them all together. For example, the product of 2, 3, 4 is $2 \times 3 \times 4 = 24$.
Proof	Given any precise logical statement, a **proof** of that statement is a sequence of logically correct steps which shows that the statement is true.
QED	**QED** is often written to mark the end of a proof. Q, E, and D are the initial letters of the latin clause *Quod Erat Demonstrandum* – meaning 'Which is what was to be proved'.
Quadrilateral	A **quadrilateral** is a polygon with four sides.
Random	When we call an event **random** it has to be the result of a process which can be repeated over and over again (such as tossing a coin, or choosing a number from a given collection of numbers). The word random then describes not just one particular event or outcome, but rather the unpredictable nature of the underlying process. The process is random if, when we use it over and over again to generate a sequence of outcomes, this sequence displays no regularity of any kind. So no matter how much one knows about the outcomes which have occurred up to some point, one cannot use this information to predict individual future events in the sequence.

Ratio	The **ratio** of two quantities is given by a pair of numbers separated by a colon which indicates the relative size of the two quantities. For example, the number of white squares and black squares in the diagram are in the **ratio** 3 : 2. □ ■ ■ □ □
Reflection symmetry	A shape has **reflection symmetry** with mirror line *m* if, when you swap the two parts on either side of the (dotted) mirror line, the whole shape lands up exactly on top of itself.
Remainder	10 goes into 23 twice, with **remainder** 3.
Rhombus	A **rhombus** a quadrilateral with all four sides of equal length.
Rotational symmetry	A rectangle has **rotational symmetry** about its centre *P* (because it lands up exactly on top of itself when we rotate it through $180°$ about the point *P*).
	A shape has **rotational symmetry** (or point symmetry) about a point *P* if it is possible to map it exactly on top of itself by rotating it through some angle less than $360°$ about *P*.
Scale factor	When a figure is enlarged with **scale factor** *k*, each line segment in the enlargement is *k* times as long as the corresponding segment in the original figure (where *k* is some fixed constant).
Segment	The line **segment** *AB* is that part of the infinite line through the two points *A* and *B* which lies between *A* and *B* (together with the two endpoints *A* and *B*).
Similar	Two shapes are **similar** if they can be labelled so that each segment of one shape is *k* times as long as the corresponding segment in the other shape, for some fixed *k*. So two shapes are similar if they have exactly the same shape, but may be of different sizes.
Square	An integer is a **square** (sometimes called a *perfect square*) if it is equal to the product of some integer by itself; (e.g. $0 = 0 \times 0, 1 = 1 \times 1, 4 = 2 \times 2$). So an integer is a square if it is equal to the area of a geometrical square with integer length sides.
Square root	$9 = 3 \times 3$ is the *square* of 3; 3 is the **square root** of $9 = 3 \times 3$. We use the symbol $\sqrt{\ }$, and write $3 = \sqrt{9}$.
Sum	The **sum** of 7 and 16 is $7 + 16 = 23$.
Symmetrical	A plane shape is **symmetrical** if it has either rotational or reflection symmetry.
Triangular number	A **triangular number** is the number of dots in a triangular array like the one shown here. Thus each triangular number is given by a sum of the form $1 + 2 + 3 + 4 + \ldots + n$ for some *n*.
Vertex	A **vertex** of a polygon is a point where two adjacent edges meet. A vertex of a three-dimensional solid is a point where three or more edges meet.